Equal Opportunity and the Case for State Sponsored Ectogenesis

Other Palgrave Pivot Titles in Philosophy

Lou Agosta: **A Rumor of Empathy**

Michael Byron: **Submission and Subjection in Leviathan**

David Coady and Richard Corry: **The Climate Change Debate**

Mihail Evans: **The Singular Politics of Derrida and Baudrillard**

Michael Hauskeller: **Sex and the Posthuman Condition**

Daniel Hill and Daniel Whistler: **The Right to Wear Religious Symbols**

Graham Oppy: **The Best Argument against God**

Graham Oppy: **Reinventing Philosophy of Religion**

George Pattison: **Paul Tillich's Philosophical Theology**

Sami Pihlström: **Taking Evil Seriously**

Barry Stocker: **Kierkegaard on Politics**

Ulrike Vieten: **Revisiting Iris Marion Young on Normalisation, Inclusion and Democracy**

palgrave▶pivot

Equal Opportunity and the Case for State Sponsored Ectogenesis

Evie Kendal
Monash University, Australia

Evie Kendal © 2015

All rights reserved. No reproduction, copy or transmission of this publication may be made without written permission.

No portion of this publication may be reproduced, copied or transmitted save with written permission or in accordance with the provisions of the Copyright, Designs and Patents Act 1988, or under the terms of any licence permitting limited copying issued by the Copyright Licensing Agency, Saffron House, 6–10 Kirby Street, London EC1N 8TS.

Any person who does any unauthorized act in relation to this publication may be liable to criminal prosecution and civil claims for damages.

The author has asserted her right to be identified as the author of this work in accordance with the Copyright, Designs and Patents Act 1988.

First published 2015 by
PALGRAVE MACMILLAN

Palgrave Macmillan in the UK is an imprint of Macmillan Publishers Limited, registered in England, company number 785998, of Houndmills, Basingstoke, Hampshire RG21 6XS.

Palgrave Macmillan in the US is a division of St Martin's Press LLC, 175 Fifth Avenue, New York, NY 10010.

Palgrave Macmillan is the global academic imprint of the above companies and has companies and representatives throughout the world.

Palgrave® and Macmillan® are registered trademarks in the United States, the United Kingdom, Europe and other countries.

ISBN: 978-1-137-54988-4 EPUB
ISBN: 978-1-137-54987-7 PDF
ISBN: 978-1-137-54986-0 Hardback

A catalogue record for this book is available from the British Library.

A catalog record for this book is available from the Library of Congress.

www.palgrave.com/pivot

DOI: 10.1057/9781137549877

Contents

Acknowledgements	vi
Introduction: The Need for Ectogenesis	1
Background: The Story Thus Far	26
1 Promoting Equal Opportunity through Ectogenesis	43
2 Protecting Equal Opportunity from Ectogenesis	62
3 Providing Equal Opportunity to Ectogenesis	93
Conclusion	112
Bibliography	115
Index	132

Acknowledgements

First I would like to acknowledge Catherine Mills, Justin Oakley and Michael Selgelid from Monash University's Centre for Human Bioethics for providing feedback on this research, as well as the anonymous peer-reviewers whose suggestions were very helpful. Thanks are also due to Erin Zillmann for providing editorial assistance, and to all the members of the Sídhe Literary Collective for supporting my academic career thus far. Finally, I wish to thank my husband and personal librarian, Zachary, for his valuable input and continued love and support.

Introduction: The Need for Ectogenesis

Abstract: *After introducing the concept of ectogenesis (artificial wombs), Kendal explores various arguments in favour of developing this technology from a liberal feminist perspective. The injustice of the unequal distribution of the physical, social and financial risks and burdens of human procreation is considered in the light of pronatalist dogma, which aims to encourage all women to become mothers, despite the numerous disadvantages associated with this choice. It is concluded that the option of ectogenesis is a necessary requirement for sexual equality in reproductive endeavours.*

Keywords: artificial womb; ectogenesis; liberal feminism; maternal morbidity; maternal mortality; pronatalism

Kendal, Evie. *Equal Opportunity and the Case for State Sponsored Ectogenesis*. Basingstoke: Palgrave Macmillan, 2015. DOI: 10.1057/9781137549877.0003.

This book will examine the case for state sponsored ectogenesis from a liberal feminist perspective, paying particular attention to issues of equal opportunity and accessibility for women. It will be using the Australian healthcare system as a case study for the introduction of this emerging technology; however, the arguments addressed will have immediate relevance to other countries that possess mixed public and private healthcare systems. Ectogenesis is commonly defined as the 'extrauterine gestation of human fetuses from conception to "birth",' although it could also entail artificial incubation of an embryo or fetus transferred from a woman's uterus after conception.[1] According to Leslie Cannold the use of assisted reproductive technologies (ARTs), such as *in vitro* fertilisation (IVF), combined with the ever-decreasing age of viability for premature infants, indicates that ectogenesis is already a 'limited reality' in many industrialised countries, including Australia.[2] As it is predicted that, Australia will experience significant economic strain resulting from its ageing population, and with infertility rates on the rise, research into new ARTs is expected to continue. Despite certain advantages for women's reproductive liberty, feminist groups remain divided regarding the desirability of current and future ARTs, including ectogenesis.[3] In this book I will be exploring one feminist perspective, arguing that the current necessity for biological gestation and childbirth impose unjust burdens on women physically, socially and financially. This injustice stems from an unequal distribution of the risks and benefits associated with human reproduction, weighted to women's disadvantage. As a result, it is my contention that the ideals of equal opportunity demand continued research into ectogenesis, such that in the future women might have the option of being liberated from these burdens when seeking to start a family.

This Introduction will discuss the nature of the physical, social and financial burdens imposed on women due to their reproductive capacity and establish the necessity of overcoming them in a way that promotes women's health and equality. As such, surrogacy is not considered a viable alternative for individual women who do not wish to undergo a physical pregnancy, as this simply shifts the burdens to other women rather than eliminating them. I will conclude this Introduction with some arguments in favour of investing in ectogenesis research, particularly in Australia.

Physical burdens associated with pregnancy and childbirth

According to Kimberley F. Curtis, 'the most central condition for women *qua* women is the biological capacity for gestation and childbirth.'[4] However, pregnancy and childbirth are known to pose numerous health risks, with some 'normal' pregnancy-related symptoms including morning sickness, dizziness, headaches, bone and muscle aches, loss of visual acuity, bleeding gums, breathlessness, heartburn, varicose veins and haemorrhoids. Importantly, despite the significant discomfort many of these symptoms impose, as they are considered 'normal' they often fail to be acknowledged seriously and many are left untreated. More concerning are the instances of gestational diabetes and preeclampsia that affect many pregnant women, not to mention the frequency of vaginal tearing and psychological trauma resulting from childbirth itself.[5] Although reliable statistics are difficult to obtain, the World Health Organisation estimates that approximately 15 per cent of all pregnant women will develop a 'potentially life-threatening complication' as a direct result of their reproductive enterprise.[6] In 'The Moral Imperative for Ectogenesis,' Anna Smajdor argues that these issues alone mean 'the claim of women to be relieved from this means of reproduction can be firmly located within a recognizably health-oriented need.'[7] She further emphasises the injustice that it is women alone who must face the physical risks associated with pregnancy and childbirth, while society at large benefits from their 'sacrifice.'[8]

Although Australia boasts one of the lowest maternal mortality rates in the world, there are still 7 reported maternal deaths per 100,000 live births in Australia,[9] with rates among Indigenous women estimated to be over five times the national average.[10] Indigenous women are also identified as being at particular risk of certain pregnancy-related complications due to the high prevalence of diabetes within this population group. Other women for whom pregnancy may represent above-average risk are those suffering kidney disease, hypertension, autoimmune disorders, chronic infections (including HIV) and transplant recipients.[11] However, even in the absence of these underlying conditions pregnancy itself represents a significant physical strain, and although the risk of maternal death is relatively small in Australia it is not non-existent.

DOI: 10.1057/9781137549877.0003

A recent case in Australia demonstrates the very real risks still faced by childbearing women, even those with access to high-quality hospital care. On 16 October 2014, 26-year-old Queensland woman Kymberlie Shepherd died in childbirth from a rare amniotic fluid embolism (AFE), a complication that can neither be predicted nor prevented.[12] Frida Simonstein and Michal Mashiach-Eizenberg claim it is remarkable that despite the fact that pregnancy 'can be deadly,' it is still not classified as an illness, suggesting this is because 'reproductive hazards have traditionally been viewed as women's fate and, therefore, taken for granted.'[13] Although improvements in maternal healthcare are certainly indicated, no level of improvement can completely eliminate the potential for at least some serious, and even fatal, complications.

Even without additional complications, pregnancy and childbirth still carry the *certainty* of some degree of physical illness and injury, with morning sickness, stretch marks and stress incontinence being among the most common. According to Maureen Sander-Staudt, natural pregnancy and childbirth have both 'debilitating' and 'disfiguring' effects on women, many of which are expected and unavoidable.[14] Under usual circumstances it would be considered only logical for someone to actively avoid developing a physical condition that is guaranteed to cause significant, prolonged discomfort, especially if it also carries the risk, however small, of sustaining severe injury or even death. Any failure to do so could be interpreted as antithetical to self-preservation. How a woman copes with the pains of pregnancy and labour, however, is treated very differently to other sources of pain – it is not typical to ask anyone if they had a 'natural, pain relief free' bout of appendicitis! While there may be some validity to this distinction (childbirth is not a disease, whereas appendicitis is) it should be noted that pain is a subjective experience and is what it is to the woman experiencing it. For some, labour can be an empowering experience, while for others it can represent intolerable suffering. Although it would seem unreasonable to expect someone suffering the debilitating cramps associated with appendicitis to forego pain relief, women still report feelings of guilt when requesting analgesia during labour.[15] According to one report in *The Daily Mail* one in four women report not receiving adequate pain relief during delivery, leading to unnecessary physical and emotional suffering. This caused some women to abandon their future family plans for fear of being denied effective pain relief during childbirth, including through unexpected epidural block failure.[16] A recent study for the *British Journal of*

Anaesthesia supports the legitimacy of this fear, citing an overall epidural failure rate of 24 per cent for caesarean sections.[17] At present, abstaining from pregnancy and childbirth is the only guaranteed way to avoid pregnancy discomfort, labour pains and birth trauma. Ectogenesis represents an alternative method of gestating a fetus that completely avoids causing injury and illness to its mother, while also avoiding the need for women to suffer through the pain of labour.

One major problem in dealing with the physical risks of pregnancy and childbirth is that, barring examples of identified high-risk groups of women, it is often impossible to predict which women will suffer the more extreme complications and negative outcomes, AFE being just one example. While this does not necessarily undermine a woman's 'freedom' to choose pregnancy, it may call into question the ability of any potential mother to make a truly informed decision to accept these risks when seeking to become pregnant. This is particularly evident when considering that these risks are so variable between women and even between different pregnancies involving the same woman. According to Amy O'Boyle et al., informed consent procedures regarding natural childbirth are in need of substantial revision, particularly with regard to the potentially devastating effects of urinary and fecal incontinence and pelvic organ prolapse following vaginal delivery.[18] They note similarities between these risks and those associated with undergoing a hysterectomy, claiming, 'Failure to include a discussion regarding the potential for urinary tract injury before the performance of a hysterectomy would be unethical. Failure to communicate the risk beforehand, document the discussion, and recognize and treat it in a timely manner would be malpractice.'[19] However, they note no comparable discussion or documentation is completed for pregnant women considering vaginal delivery, whose condition makes them particularly susceptible to sphincter damage that may be permanent (these authors relay figures as high as 35 per cent of first-time mothers and 40 per cent of multiparous women showing signs of sphincter damage as a result of childbirth).[20] I argue that this inconsistency with regard to consent illustrates how women's risks in childbearing are taken for granted.

Pain and injury are not the only physical burdens associated with pregnancy, as women are also expected to drastically alter their diet and exercise routines while pregnant. Such changes in behaviour are often advised to begin months before conception and continue through gestation and lactation, which together represent a significant time commitment. For some women the recommended dietary and lifestyle changes can be particularly burdensome, especially for those who smoke or drink alcohol, and those with a preference for foods that increase the risk of developing certain infections that are known to cause fetal damage. Examples include listeriosis, caused by ingestion of contaminated meat and dairy products, with soft cheeses representing a particularly high-risk food; trichinosis, most commonly caused by ingestion of undercooked pork; and toxoplasmosis, a parasitic infection caused by ingestion of contaminated meat products, or handling of feline faeces.[21] In addition to harming the fetus, for all three infections pregnancy significantly increases the risk of life-threatening complications for the woman.

Women whose occupations or leisure activities involve excessive physical exertion are also likely to have to sacrifice these interests or work commitments while pregnant. Tim Bayne and Avery Kolers remind us that a fetus is 'materially derived' from its gestational mother, stating 'inside her, the embryo-foetus is actually created and, at significant expense, she provides the oxygen, nutrients and shelter required to bring the foetus to term.'[22] This also highlights the fact that the woman loses a degree of bodily integrity while pregnant, as another human life now occupies the space inside her own body. As a result, freedom of movement is severely impeded during pregnancy not only due to increased physical size, but also the fact that pregnant women are advised to avoid certain environmental exposures and often cannot receive travel vaccinations.[23] The impact such restrictions will have on each individual woman will depend on various lifestyle factors, but could be expected to range from trivial inconvenience to severe disruption of everyday activity and employment.

Pregnancy also greatly restricts the freedom of medical treatment for both the mother and the fetus, precisely because the fetus exists *inside* the body of its gestational mother. A pregnant woman's immuno-compromised state makes her particularly susceptible to certain illnesses, including many that cannot be treated effectively without risking harm to the fetus. This can either be because the drugs required to treat the condition are known to be teratogenic, or because there is insufficient information available to suggest they are not, since pregnant women are generally excluded from the drug trials required to make this assessment.[24] The same dilemma is present for women who discover they have cancer while pregnant, as current cancer treatments are potentially very harmful to the fetus. This is one area in which I argue ectogenesis could make a particular contribution to women's welfare, as women would no longer have to choose between their health and that of their fetus when considering chemotherapy and radiation treatments – they could choose instead to transfer their fetus to an artificial womb before commencing treatment.

Similar arguments can be made in favour of ectogenesis allowing easier access to the fetus if surgical intervention were required, for example, to correct a neural tube defect. Research has indicated that the disabling effects of spina bifida could be minimised by performing fetal surgery earlier in gestation than is currently possible through intrauterine methods.[25] Likewise, intrauterine blood transfusion for Rhesus blood type incompatibility, in which the blood of the fetus may be attacked by maternal antibodies, could be replaced by incubating the embryo or fetus ectogenetically, thus avoiding haemolytic anaemia of the newborn and minimising the distress for the mother caused by repeated invasive procedures.[26] It is also likely that more fetal interventions could be successfully performed if the fetus could be treated independent of its mother, and earlier diagnosis of potentially debilitating conditions made possible through the enhanced visibility of an extrauterine fetus.

Ectogenesis would not only eliminate the pain of childbirth and possibility of birth injury to both the mother and fetus, as no birth event takes place, but also has the potential to prevent much of the physical damage sustained through natural pregnancy. Growing the fetus outside the mother's womb would also prevent feto-maternal cell transfer, and the retention of 'foreign' fetal cells after pregnancy, which has been implicated in the development of some autoimmune disorders.[27] Similarly, fetuses could be protected from harmful substances being delivered through the placenta from their mothers, including certain maternal antibodies and environmental toxins. Depending on how ectogenesis is achieved in the future, and particularly whether *in vitro* initiation or embryo or fetal transfer is used, it could also circumvent the low probability of maternal death and replace it with zero risk of mortality.

Resituating gestation outside the mother's body could also alleviate a lot of the physical constraints currently imposed on pregnant women, including those that can compromise their own medical care, by restricting what medications and treatments they can receive, and require them to undergo surgical procedures intended to treat the fetus. Remembering that with the exception of living organ donors, it is rare for any patient to be subjected to the risks of surgical intervention in order to treat a *different* patient; however, caesarean sections represent an example of a major abdominal surgery performed on the mother for the sake of the fetus. Caesareans are being performed ever more frequently in industrialised countries like Australia, but numbers could decline if some high-risk pregnancies were replaced with artificial gestation. While many of the physical burdens of pregnancy and childbirth could be removed through partial ectogenesis, others, such as preventing the development of autoimmune disorders caused by feto-maternal microchimeras, would depend on full ectogenesis. Gestational age would also determine whether a fetus would need to be surgically extracted before treating it or its mother for some of the conditions listed earlier, which would carry its own risks. Given the fact that currently no alternative exists for women who, for example, need to undergo chemotherapy but do not wish to harm their fetus, fetal extraction and subsequent artificial incubation may be the preferred choice for some pregnant women in this situation.

Regardless of how an individual woman may feel toward both the normal physiological changes of pregnancy and the possibility of more pathological ones, the fact remains that for every woman there *are* physical changes associated with getting pregnant and giving birth, and many of these can have a serious impact on future health. Shulamith Firestone declares in her famous 1970s text, *The Dialectic of Sex: The Case for Feminist Revolution*: 'Pregnancy is barbaric.... Pregnancy is the temporary deformation of the body of the individual for the sake of the species. Moreover, childbirth *hurts*. And it isn't good for you.'[28] Firestone argues that women's oppression is inextricably linked to their biological capacity for childbearing and that equal opportunity demands that alternative methods for propagating the human species, such as ectogenesis, are developed as soon as possible to redress this inequality. While parts of Firestone's radical argument centre on the abolition of natural pregnancy as a necessary step toward achieving equality, in this book I am defending the more moderate position she proposes, in which ectogenesis represents a much-needed *option* for those women who wish to have children without submitting to the physical burdens of gestation and childbirth. Promoting new reproductive alternatives for women who would benefit from them does not require that existing methods be denigrated or undermined. Creating a family is a very personal experience and there will never be a one-size-fits-all solution to the issues raised concerning women's reproductive liberty and choice. What I hope to demonstrate throughout this book, however, is that ectogenesis offers some unique advantages that are not provided by existing options.

Social burdens associated with pregnancy and childbirth

In addition to the physical risks of pregnancy and childbirth there are also social burdens associated with women's reproductive capacity. Perhaps chief among these is the influence of potentially coercive pronatalist agendas, aimed at encouraging women to postpone or entirely abandon other life goals in order to devote themselves to achieving the socially condoned 'ideal' of motherhood.[29] Dorothy E. Roberts claims that such social pressures are often first exerted on women to *encourage* pregnancy, and then used as a justification to socially police the behaviour of women *while* pregnant. According to Roberts this social aspect of pregnancy can be particularly oppressive to women, leaving them open to various abuses, including medical paternalism, emotional manipulation and a lack of respect for their individual privacy and autonomy.[30] This section will explore some of these claims.

Eileen Fischer et al. define pronatalism as 'the belief that people should have children, regardless of the means required to become parents,' claiming this belief owes its origins to certain 'cultural assumptions' about the normalcy of parenthood and the aberrance of childlessness.[31] While these authors focus predominantly on industrialised societies in their analysis, they claim that to some extent *all* cultures 'valorize parenthood,' often in ways deeply rooted in their unique socio-historical context.[32] Australia has been identified as a pronatalist society, which from a feminist perspective suggests the educational, social and occupational development of individual Australian women may be adversely affected by the privileging of children as an invaluable source of cultural capital. In the pronatalist belief system the absence of children correlates with significant social disadvantage, as much of society is structured around the concept of the biological family unit. While pronatalism is often defended on the grounds of encouraging population growth, particularly in those nations with an ageing population, it nevertheless creates an expectation for women to submit to pregnancy and childbirth as a means of attaining social status, a decision which may materially damage their other life goals and aspirations.

According to Alena Heitlinger, pronatalism serves to promote the 'motherhood mandate,' by portraying motherhood as 'normal' and 'central to a woman's identity,' thus manufacturing a universality that fails to treat women as individuals.[33] Similarly, Diana Meyers asserts that in a pronatalist society women are socialised to consider childbirth an 'inevitable part of life,' presumably even when contraceptives are readily available to avoid pregnancy.[34] Thus, due to the influence of pronatalist discourse, many women may pursue motherhood without ever truly considering the alternatives, going along with cultural indoctrination rather than making informed, autonomous reproductive choices. Furthermore, Meyers blames pronatalistic social coercion for the fact that when women experience difficulty in conceiving, 'they often display a monomaniacal dedication to infertility treatment.'[35] As pronatalism equates childlessness with abnormality, it exerts undue pressure on infertile women to volunteer for risky, invasive and often unsuccessful medical interventions aimed at achieving pregnancy. However, given the physical burdens of pregnancies conceived naturally, I further argue that pronatalistic social coercion also leaves fertile women vulnerable to exploitation, by promoting the idea that women's fulfilment is inextricably bound to their childbearing capacity. That women are often socialised from a young age to pursue self-actualisation through procreation can therefore be considered a form of social oppression, whether or not an individual woman actually chooses to have children, and regardless of any social benefit they may receive if they do.

One of the major goals of pronatalism is to encourage childbearing by advancing a romanticised view of motherhood as being necessarily linked to womanhood. This leads to what Meyers refers to as the 'compulsion' for women to procreate, despite an inability to 'accurately anticipate and fully appreciate the consequences of this choice.'[36] This is true regardless of any previous experience of pregnancy or parenthood, as each pregnancy and its resultant offspring is unique. Thus, pronatalistic enculturation may diminish women's autonomy in their reproductive decisions, leaving them subject to overwhelming social pressures and 'unconscious desires.'[37] This is achieved both through cultural indoctrination regarding the desirability and normalcy of procreation, and attempted concealment of the negative aspects of pregnancy, childbirth and parenthood. The net result of these influences aims to prevent women seriously considering the alternatives to having children, even when significant obstacles need to be overcome to make this a possibility, as in the case of infertility. Ann Oakley notes that despite the liberating effects of the contraceptive revolution, the pressures of social expectation mean 'few women feel they really can avoid becoming mothers.'[38] This fact is particularly important when considering the issue of women's autonomy in choosing to accept the risks and burdens of pregnancy and childbirth, as in a pronatalist society some reproductive decisions are likely to be motivated by a desire to avoid the stigma of infertility and social disapproval of voluntary childlessness. Importantly, all women in such societies are subjected to the social pressures associated with pronatalist dogma, regardless of their individual willingness or ability to conceive. Women are thus left with the option of either conforming to social pressure and pursuing motherhood, even at the expense of other life goals, or deviating from societal norms and risking public censure and social exclusion for choosing not to procreate. Unlike the physical burdens of pregnancy, pronatalist discourse is a burden associated with pregnancy that exerts its influence on women irrespective of whether or not they ever actually become pregnant.

Although pronatalistic social coercion also influences men's reproductive decisions, I argue it disproportionately affects women by glorifying motherhood and viewing infertile women as worthy of pity, while also portraying voluntarily childless women as deviant, selfish or immature. To illustrate the difference one only needs to consider the standard questions men and women face in new social situations. While a positive answer to 'Do you have children?' certainly adds status for a man, a negative answer is not generally met with a demand for explanation or considered to demonstrate a significant lack, assuming he has satisfactorily answered the primary question: 'What do you do?' Fatherhood is thus an agreeable 'extra' but not considered as fundamental to his masculine identity as his occupation. This attitude does a disservice both to stay-at-home dads and men not currently in paid employment. For a woman the hierarchy of these questions, if not their order, is likely to be reversed, with childlessness, especially due to the demands of a career, remaining stigmatised. Obviously, there are still various negative stereotypes regarding parenting roles and voluntary childlessness that would need to be abandoned in order to promote autonomous decision making among women and couples considering parenthood.

Research has shown that childless couples are often made to feel socially inferior to parents, despite the fact that they are likely to enjoy higher levels of education, greater occupational satisfaction and better couple communication and interaction than their childrearing counterparts.[39] In a pronatalist society social education regarding family life also neglects to inform potential parents of the true impact having children has on personal relationships, society and the environment, and instead promotes an ideology in which procreation is the norm, and childlessness requires justification.[40] As such, I suggest that the social burden of pregnancy is twofold: social expectation demands women become pregnant, and then pregnancy serves to materially damage women's social lives and limit their future opportunities. As Bayne and Kolers point out, 'The social norms around childbirth and rearing are among the more regressive, with respect to gender, in our society,' with women suffering significant social disadvantage due to their reproductive biology.[41] This is a result of the combination of the physical constraints of pregnancy limiting women's social and employment activity during this time, and the social policing of pregnant women's bodies imposing restrictions on their behaviour and lifestyle choices. Examples of such policing are numerous, but perhaps most obviously include social disapprobation for women who smoke, drink alcohol or sometimes even those who simply choose to continue paid employment while pregnant.

Another social burden of pregnancy and childbirth that derives from these relates to the loss of personal privacy many women experience while pregnant. While much of medical ethics is focused on maintaining patient privacy and confidentiality, pregnancy represents one example in which women's health is under constant public scrutiny. According to Madison Powers, autonomy requires the ability to 'make important life choices free from the scrutiny of others,' which I argue is not possible when the life choice at stake – becoming pregnant – is so publicly visible.[42] As it is not generally possible to conceal the latter stages of pregnancy (or the rather more conspicuous resultant offspring), privacy regarding their reproductive decisions is automatically reduced for women, whose bodies quite literally declare their reproductive status in a way that men's bodies simply do not. Sander-Staudt also notes that 'the visible signs of pregnancy often give others a feeling of entitlement to offer unsolicited advice or to touch a woman's body,' directly compromising that woman's right to privacy and exposing her to rigorous social policing.[43] In this instance, ectogenesis represents the only viable means of completely

Introduction: The Need for Ectogenesis 13

protecting a woman's privacy, and by extension her autonomy, as a woman could be 'expecting' a child while still controlling who has access to this private information. With regard to gender equality, men are currently able to conceal a lot more about their reproductive endeavours than women, and can also procreate without risking infringements on their personal space or medical privacy. They are also exempt from social policing regarding diet and exercise when procreating and do not suffer the same physical restrictions imposed by gestation that can severely impact a woman's employment and leisure. This is in spite of increasing evidence that sperm quality is impacted by lifestyle factors, including exposure to tobacco, alcohol, drugs and workplace chemicals and pesticides.[44] While the other impacts of pronatalism may not be directly influenced by the advent of ectogenesis technology, enhancing privacy and personal liberty regarding reproductive decisions represents a significant advantage for women. Ectogenesis would also challenge various stereotypes about parenting roles and provide one possible solution to the perceived conflict between women's identities as independent agents and as potential incubators.

Pronatalism has developed to promote population growth, which heretofore has depended entirely on women gestating the next generation. The potential role of ectogenesis in this belief system is complex and may on the surface appear to be contradictory. While ectogenesis would challenge the assumption that women have to sacrifice other interests to be pregnant, it could also be seen as capitulating to pronatalism by allowing more women to satisfy the demands of this cultural indoctrination that might otherwise be unable to. I view this issue merely as a conflict between short- and long-term benefits. In the short-term ectogenesis may well be usurped to serve a pronatalist agenda by providing yet another means of promoting the 'motherhood mandate,' but in so doing it would also alleviate some of the suffering of women already living under this system, particularly those with fertility problems. Importantly, reproductive biotechnologies have not caused pronatalist dogma, but they do have the power to reinforce it. In the long term, ectogenesis has the potential to challenge the very foundation of pronatalism, by providing an image of the future in which population growth is not solely dependent on enticing, manipulating, or coercing women to be pregnant.

From an equal opportunity standpoint it is significant that while both male and female parents share the benefits a child confers in a

DOI: 10.1057/9781137549877.0003

pronatalist society, it is only the woman who is subjected to the *direct* physical and social risks associated with bringing that child into existence. This is not to say there are no concomitant social benefits that may be specific to gestational mothers, but rather that where there is social disadvantage accompanying the birth of a child this disproportionately affects the mother. Debarun Majumdar claims this is mostly due to the fact that following the birth of a child 'the mother usually reverts back to her traditional role,' bearing most of the responsibility for domestic duties often despite any prior arrangement to share such responsibilities equally with their partner.[45] Overall, as Majumdar states, in a pronatalist society 'males have less to lose and more to gain socially than females in the event of a birth.'[46] I argue that a major contributing factor for this is the difference having children has on employment opportunities for men and women, with women's careers often being more adversely affected.

Financial burdens associated with pregnancy and childbirth

In addition to the physical and social implications of pregnancy and childbirth for women, there are various economic disadvantages that women face if they wish to procreate. These include the necessity of temporary withdrawal from paid employment in order to give birth and the potential for prolonged absences for childrearing purposes. Linda R. Hirshman notes that any length of time spent away from work impacts a woman's financial security and independence and can materially damage her future earning capacity due to a perceived loss of 'human capital' resulting from her leaving the workforce, however briefly.[47] While she admits that devoting time to raising children has 'obvious emotional and immediate rewards,' she argues domestic life provides fewer opportunities for 'full human flourishing' than exist in the public domain.[48] For Hirshman, the fact that following the birth of a child many women are relegated to this 'less flourishing sphere' constitutes a major injustice, even in cases where women assign themselves the 'repetitious, socially invisible' task of raising a family.[49] Although most of the financial disadvantage women face due to childbearing is connected with the care of children rather than pregnancy itself, I argue the physical incapacitation of gestation lays down the foundation for gender inequality in childrearing responsibilities.[50] As it is necessarily the woman who 'slows down' and takes time away from paid employment in order to be pregnant, this makes it easier for them to slip uncritically into domesticity after childbirth, perpetuating an unequal distribution of the burdens of childrearing. While a lot of social changes are required to combat this problem, I argue technological alternatives to physical gestation represent one method of challenging this norm.

Once women leave the workforce to give birth, they are likely to experience significant difficulty in breaking free from the confines of the home to return to their former positions. This is particularly true when their return to paid employment would necessitate full-time childcare, as the expense of this is often calculated as exclusively coming out of the woman's salary. Hirshman notes that when returning to work would cost a significant portion of the woman's potential earnings in childcare it is often assumed that she should stay home to save money, despite the fact there is no reason to mentally deduct the cost of childcare from the woman's income alone, rather than the household income.[51] Furthermore, this attitude assumes women have no other reason to return to the public sphere after childbearing than immediate income generation, thereby denying the fact that returning to work sooner could have long-term benefits for women's careers and social lives beyond this initial period, where the cost of childcare may well make the return to work less financially lucrative. Other obstacles women face when attempting to re-enter the workforce include hostility from co-workers and supervisors when returning part time and a lack of affordable childcare facilities. All of these obstacles need to be addressed in order to minimise the financial disadvantage many women face due to their reproductive choices.

When discussing the unequal number of men and women in positions of social authority, Oakley notes that it is childbearing for women that is 'more likely to bring downward occupational mobility than anything else.'[52] She claims that all other things being equal, a male and female of comparable experience and qualifications may begin on the same career path with similar prospects, but time spent away from paid employment due to family responsibilities will soon leave the woman lagging behind, both in terms of pay and position. Pregnancy and childbirth can represent a severe impediment to career advancement for women, since a combination of biological necessity and social convention demands that women should aim to start a family at a time in life that often coincides with a vital stage of their professional development. Ectogenesis would make it possible for these women to continue their careers uninterrupted, thereby encouraging equal opportunity for promotions and equitable distribution of the benefits of continuous service. At present, in cases where taking time out from paid employment to be the primary caregiver of a new child would have less impact on a male partner's career, the necessity of biological gestation still requires that the female partner sacrifice some work time for pregnancy-related illness and birthing. This is not the case for the reverse scenario. While I would argue that neither parent *should* eschew all time-consuming responsibility for their offspring, it is still important to consider that biological difference makes it such that only one parent *can*. The impact of this on social attitudes towards women's 'natural' role as primary caregivers is impossible to quantify; however, it is likely to affect all women in society to some degree, including the voluntarily childless. For those women who do intend to be the primary caregivers of their children, ectogenesis would make it possible to schedule childbearing at whatever stage of life is most convenient, even if this is beyond the ideal reproductive years.

Although not likely to be directly impacted by the potential for ectogenesis, the disproportionate amount of unpaid domestic labour performed by women may also involve economic exploitation, particularly as many of the benefits of employment are not accrued by stay-at-home mothers or housewives. Also, while their partners may financially support many of these women during the initial childrearing period, Nancy Folbre notes that the rise in divorce makes it 'economically risky' to be a housewife in the long term.[53] Regardless of any immediate financial support provided by social welfare or a woman's domestic partner, time lost in employment for childrearing responsibilities often cannot be compensated for later, with missed opportunities and promotions likely to have lasting effects on women's economic potential.

In *Pregnant Men*, Ruth Colker argues that the mere *ability* to become pregnant 'has historically been an excuse for denying women equal employment opportunity,' regardless of whether or not an individual woman intends to become pregnant.[54] She identifies this discriminatory treatment of women on the basis of biological difference as a clear example of sex discrimination, demonstrating the economic disadvantage associated with women's reproductive capacity. In addition to cases of open discrimination, where women are specifically targeted for exclusion from certain occupations based on perceived risks to their reproductive health or the potential for fetal damage, there are many instances of hidden discrimination in which women are intentionally 'overlooked' for certain positions, in order to avoid the hassle and expense of arranging maternity leave should they decide to take time off to have children. The more senior the position, the more likely it will be difficult to find a short-term replacement to accommodate leave for family responsibilities, thus providing an incentive for companies to hire men in these positions. Although such discrimination is illegal in countries like Australia, it could be difficult to prove, for example, that the real reason a female job applicant was rejected was that she turned up to the interview heavily pregnant. Again, the visibility of pregnancy is to blame for this disadvantage, as men can conceal their family plans in ways women cannot. Since there is currently no other way to gestate human fetuses, employers may believe, rightly or wrongly, that they can safely avoid the demands of providing parental leave, simply by excluding women from their workforce.

While the Australian government's new paid parental leave scheme allows either parent to assume primary care of a child after birth, the fact that women need to gestate the fetus inside their body means they will still require flexible leave arrangements in order to give birth.[55] As noted earlier, once women take time out of employment for childbearing they are much more likely to find themselves the primary carer of the resultant child, despite any prior agreement with their partner to share childrearing responsibilities. The persistence of the 'wage gap,' in which women are paid less to do the same jobs as men, also diminishes women's negotiating power when it comes to deciding which parent will stay home to care for the children, thereby compounding financial disadvantage for women as a result of sex-based discrimination. I argue this leads to a vicious cycle in which women are handicapped in competition for employment and promotion due to the increased likelihood that they will be the ones to assume primary care of their children, with this likelihood being further increased by the fact such disadvantages in employment opportunity mean many women are earning less than their male counterparts. Ectogenesis represents the only method of procreating that would impose no additional employment restrictions on women compared with men, as both could continue working throughout the entire gestation period, regardless of the type of employment involved. Also, I suggest that if women were not already incapacitated by the physical requirements of pregnancy and childbirth, the probability that they would slip uncritically into the role of homemaker after giving birth would decrease, especially if there was an accompanying reduction of the wage gap between men and women in employment. Such a reduction could be caused in part by attacking the assumption that women are a bad investment for promotion, by making it equally likely that men will become the primary caregivers for their children.

Ectogenesis research and why Australia should lead the way

Having discussed some of the ways in which women are specifically disadvantaged due to their capacity for pregnancy and childbirth, I now wish to explore in greater detail the potential benefits for developing ectogenesis as a technological alternative to these biological processes. In terms of practical application, it is already known that the more sophisticated artificial incubation chambers become, the more successful they will be in incubating to viability infants born increasingly premature.[56] With advances in IVF technology also extending the time an embryo can develop *ex utero* before implantation, there is now a limited window in which 'natural' pregnancy is necessary at all, with some estimates based on current technology being as low as 10–12 weeks of the total 40-week gestation.[57] As this window continues to diminish, biological gestation may soon become entirely redundant, necessitating a re-evaluation of the supposed 'intrinsic value' of pregnancy.

The physical, social and financial burdens associated with pregnancy discussed in this Introduction clearly demonstrate the potential advantages for women were ectogenesis to become a reality. Scientists currently working on developing an artificial placenta further note that the potential health benefits for ectogenesis would not be limited to the mother alone, but would include the possibility of independent fetal surgery, and the maintenance of an ideal growth medium free from the environmental hazards that the fetus may be exposed to in the uncontrolled uterine environment.[58] Exposure to teratogens while *in utero* is responsible for the development of various disabilities and birth defects. In many of these cases the damage to the fetus is sustained before the woman knows she is pregnant.[59] Although fetal, neonatal and perinatal death rates are comparatively low in Australia, such exposures, as well as various other complications during pregnancy, contribute significantly to their values.[60] Given the increased monitoring of pregnant women's behaviour, and the risk of legal action against those considered to be harming their unborn children, the significance of developing an ideal extrauterine environment for fetal incubation should not be underestimated.[61] Ectogenesis also carries the obvious benefit of saving the lives of more babies born prematurely and reducing the suffering of women whose wanted pregnancies spontaneously abort, by providing a means of continuing gestation artificially. As such, developing this technology would have both preventive and therapeutic value.

Given the strengths of the advantages listed so far, it would be reasonable to expect that funding into ectogenesis research would be a priority in countries medically advanced enough to attempt it. However, this is definitely not the case. On the contrary, legislation in many industrialised nations actively impedes research in this area of reproductive technology. Simonstein and Mashiach-Eizenberg note that most countries ban embryo experimentation beyond the 14th day of development, stating that Canada in particular has 'explicitly prohibited any research designed to add to the knowledge of ectogenesis.'[62] Similarly, in the United States no federal funds can be used towards ectogenesis research.[63] More tangentially, certain laws in the United Kingdom and Europe assign legal parenthood for women *on the basis* of gestation rather than genetics, necessarily imposing a different standard for establishing motherhood compared with fatherhood.[64] While this does not directly oppose ectogenesis research it does expose a certain attitude towards women's role in procreation that would not accommodate artificial gestation. According to Firestone, such attitudes impact the funding available for reproductive research, often exposing a 'cultural lag and sexual bias' that demands any advances in artificial incubation be 'excused' on the grounds of promoting fetal survival, rather than women's interests.[65] In these jurisdictions, as in Australia, restrictions on research involving human embryos also serve as a major impediment to ectogenesis research. Australia's *Prohibition of Human Cloning for Reproduction and the Regulation of Human Embryo Research Amendment Act* of 2006 lists as an offence 'creating an embryo for a purpose other than achieving a pregnancy in a woman,' imposing a maximum penalty of imprisonment for 15 years. The same prohibitions and penalties are imposed for 'developing a human embryo outside the body of a woman for more than 14 days' and 'collecting a viable human embryo from the body of a woman.'[66] All of these restrictions effectively prohibit research into technological alternatives to physical gestation, by criminalising all the requisite steps to achieving the goal of full human ectogenesis. I now wish to propose various reasons why instead of providing opposition, Australia should in fact be leading the way in ectogenesis research.

Australia has a proud history of advancing medical research in general, and has been at the forefront of many developments in reproductive technology in particular. The first IVF pregnancy was achieved in Australia in 1973, and although the resultant embryo survived only a few days, this still represented a significant breakthrough in infertility treatment. Following the birth of the world's first 'test-tube' baby, Louise Brown, in 1978, Australian researchers were also the first to repeat the success of Patrick Steptoe and Robert Edwards' team in the United Kingdom, leading to the birth of Candice Reed in 1980.[67] According to Peter Singer and Deane Wells, a short time later the Melbourne-based fertility clinic responsible boasted success rates that 'surpassed that of the English originators of the process.'[68] The clinic, which was run by Monash University's Professor Carl Wood, also pioneered the process of embryo donation in 1983 and achieved the first healthy delivery of a baby developed from a frozen embryo in 1984.[69] As such, there is historical precedent suggesting Australian researchers are capable of making significant contributions to assisted reproductive medicine. Although some of these advancements have met with public criticism, the majority have been accepted and incorporated into infertility care, indicating the Australian population may be willing to embrace more technologically dependent forms of reproduction, including ectogenesis. Furthermore, although the law is different from state to state, parts of Australia have abortion laws that are sufficiently liberal to potentially ease the ethical dilemmas facing researchers engaging in the earlier stages of ectogenesis research, which would undoubtedly entail significant losses.[70] That such research is more likely to be socially tolerated in a country in which abortion is legal should be fairly clear. From an equal opportunity standpoint it is also relevant that Australian society values the ideal of a 'fair go' for all, suggesting the potential for ectogenesis to enhance equality for women is likely to hold significant traction.

Another factor making Australia a suitable forerunner for ectogenesis research is Australia's unique two-tiered healthcare system, which would allow for an evaluation of the efficacy of various public and private funding options for the provision of ectogenesis services. The ethical issues surrounding inequitable provision of ARTs are often conflated with those of developing the technologies themselves. Concerns about class inequality replacing gender inequality in such a case make it clear publicly funded services are necessary to preserve social justice. Australia's public healthcare system, Medicare, already provides rebates for fertility treatment, without imposing as harsh eligibility criteria as is common to other public healthcare systems around the world.[71] This implies the Australian government may be more willing than most to entertain the idea of state sponsored ectogenesis, particularly as a means of counteracting the population decline that is currently threatening Australia's future economic stability. Recent government initiatives aimed at encouraging childbearing and increasing family sizes in Australia seem to support this analysis.

Introducing *Equal opportunity and the case for state sponsored ectogenesis*

Although there has already been some valuable research done in Australia regarding the ethical implications of ectogenesis, particularly by such authors as Cannold and Stephen Coleman, for this book I have chosen to focus on the equal opportunity grounds for supporting ectogenesis research in Australia and beyond. The Background section will cover a brief history of ARTs and the feminist and conservative arguments surrounding them. Chapter 1 will explore the need to promote equal opportunity between men and women in their reproductive pursuits, between women and other women in terms of reproductive choices, and will also briefly touch on the issue of sexual orientation and access to ARTs. The contention of this chapter will be that developing ectogenesis is a necessary condition for maximising reproductive choices for all citizens, and that this will lead to greater social and economic equality. Chapter 2 will examine the need to protect current anti-discrimination laws, particularly as they pertain to maternity leave arrangements and pregnancy-based discrimination in the workplace, should ectogenesis become a viable alternative to natural pregnancy. This chapter will also address potential challenges to abortion rights that ectogenesis could pose, the need to protect the right of women to refuse new ARTs, and the possible implications that ectogenesis might have for the concept of fetal rights and discrimination. Chapter 3 will look at the issue of funding ectogenesis, arguing for state sponsored research and distribution to ensure equal access to the technology, regardless of socio-economic status. Alternative distributive systems will also be explored, including allowing a private, user-pay system to arise or limiting public access to cases of medical necessity. I will conclude with a discussion of the major obstacles to ectogenesis research that would need to be overcome before this technology could be realised. I will adopt a liberal feminist methodological framework, while exploring some of the conflicting feminist views regarding ectogenesis as a potential solution to the unequal distribution of the burdens associated with reproduction.

Notes

1. Mary Anne Warren, 'Making Babies: The New Science and Ethics of Conception by Peter Singer; Deane Wells,' *Ethics* 97, no. 1 (1986): 288.
2. Leslie Cannold, 'Women, Ectogenesis and Ethical Theory,' *Journal of Applied Philosophy* 12, no. 1 (1995): 56.
3. Julien S. Murphy, 'Is Pregnancy Necessary? Feminist Concerns about Ectogenesis,' *Hypatia* 4, no. 3 (1989): 66.
4. Kimberley F. Curtis, 'Hannah Arendt, Feminist Theorizing, and the Debate over New Reproductive Technologies,' *Polity* 28, no. 2 (1995): 162.
5. Anna Smajdor, 'The Moral Imperative for Ectogenesis,' *Cambridge Quarterly Healthcare Ethics* 16, no. 3 (2007): 340.
6. WHO Department of Reproductive Health and Research, *Managing Complications in Pregnancy and Childbirth: A Guide for Midwives and Doctors* (Geneva: World Health Organisation, 2007), v.
7. Smajdor, 'The Moral Imperative for Ectogenesis,' 340.
8. Ibid., 336.
9. CIA World Factbook, 'Australia-Oceania,' 21 April 2015. Available at: https://www.cia.gov/library/publications/the-world-factbook/geos/as.html.
10. Australian Bureau of Statistics (ABS), 'The Health and Welfare of Australia's Aboriginal and Torres Strait Islander Peoples, 2008,' 22 May 2010. Available at: http://www.aihw.gov.au/WorkArea/DownloadAsset.aspx?id=6442458617.
11. Peter Singer and Deane Wells, *Making Babies: The New Science and Ethics of Conception* (New York: Charles Scribner's Sons, 1985), 98; Gillian Lockwood, 'Pregnancy, Autonomy and Paternalism,' *Journal of Medical Ethics* 25, no. 6 (1999): 538.
12. Patrick Williams, 'Fiancé of Kymberlie Shepherd, Who Died during Childbirth, Calls for More Research into Rare Amniotic Fluid Embolism,' *ABC News*, 31 October 2014. Available at: http://www.abc.net.au/news/2014-10-31/queensland-mother-dies-from-rare-amniotic-fluid-embolism/5844512.

13 Frida Simonstein and Michal Mashiach-Eizenberg, 'The Artificial Womb: A Pilot Study Considering People's Views on the Artificial Womb and Ectogenesis in Israel,' *Cambridge Quarterly of Healthcare Ethics* 18 (2009): 88.
14 Maureen Sander-Staudt, 'Of Machine Born? A Feminist Assessment of Ectogenesis and Artificial Wombs,' in *Ectogenesis: Artificial Womb Technology and the Future of Human Reproduction*, eds Scott Gelfand and John R. Shook (New York: Rodopi, 2006), 112.
15 Eran R. Horowitz, et al., 'Women's Attitudes toward Analgesia during Labor – A Comparison between 1995 and 2001,' *European Journal of Obstetrics & Gynecology and Reproductive Biology* 117, no. 1 (2004): 31.
16 Rachel Ellis, 'The Botched Epidurals Making Women Terrified of Giving Birth: Is This the Real Reason for Soaring Caesareans?' *The Daily Mail*, 23 November 2010. Available at: http://www.dailymail.co.uk/health/article-1332129/Botched-epidurals-making-women-terrified-giving-birth-Is-real-reason-soaring-Caesareans.html.
17 J. Hermanides, et al., 'Failed Epidural: Causes and Management,' *British Journal of Anaesthesia* 109, no. 2 (2012): 144. The authors note that incorrect catheter placement was responsible for half of these cases.
18 Amy L. O'Boyle, et al., 'Informed Consent and Birth: Protecting the Pelvic Floor and Ourselves,' *American Journal of Obstetrics and Gynecology* 187, no. 4 (2002): 981.
19 Ibid., 983.
20 Ibid.
21 Patricia Kendall, et al., 'Food Handling Behaviors of Special Importance for Pregnant Women, Infants and Young Children, the Elderly, and Immune-Compromised People,' *Journal of the American Dietetic Association* 103, no. 12 (2003): 1648.
22 Tim Bayne and Avery Kolers, 'Toward a Pluralist Account of Parenthood,' *Bioethics* 17, no. 3 (2003): 238.
23 Jonathan Cohen, 'The Pregnant Traveller,' *Medicine Today* 9, no. 5 (2008): 65.
24 Marleen M.H.J. van Gelder, et al., 'Teratogenic Mechanisms of Medical Drugs,' *Human Reproduction Update* 16, no. 4 (2010): 379.
25 Leslie Sutton, 'Fetal Surgery for Neural Tube Defects,' *Best Practice and Research Clinical Obstetrics and Gynaecology* 22, no. 1 (2008): 175.
26 Sander-Staudt, 'Of Machine Born?' 119.
27 Gavin Dawe, et al., 'Cell Migration from Baby to Mother,' *Cell Adhesion and Migration* 1, no. 1 (2007): 19.
28 Shulamith Firestone, *The Dialectic of Sex: The Case for Feminist Revolution* (New York: William Morrow and Company, 1970), 198.
29 Diana Meyers, 'The Rush to Motherhood: Pronatalist Discourse and Women's Autonomy,' *Signs* 26, no. 3 (2001): 747.

30 Dorothy E. Roberts, 'The Genetic Tie,' *The University of Chicago Law Review* 62, no. 1 (1995): 240.
31 Eileen Fischer, et al., 'Pursuing Parenthood: Integrating Cultural and Cognitive Perspectives on Persistent Goal Striving,' *Journal of Consumer Research* 34, no. 4 (2007): 428.
32 Ibid.
33 Alena Heitlinger, 'Pronatalism and Women's Equality Policies,' *European Journal of Population* 7, no. 4 (1991): 344.
34 Meyers, 'The Rush to Motherhood,' 746.
35 Ibid., 747.
36 Ibid., 752.
37 Ibid., 743.
38 Ann Oakley, 'Gender and Generation: The Life and Times of Adam and Eve,' in *Women and the Life Cycle: Transitions and Turning-Points*, eds Patricia Allatt, Teresa Keil, Alan Bryman and Bill Bytheway (Essex: Macmillan Press, 1987), 27.
39 Satoshi Kanazawa, 'Intelligence and Childlessness,' *Social Science Research* 48 (2014): 157; Harold Feldman, 'A Comparison of Intentional Parents and Intentionally Childless Couples,' *Journal of Marriage and Family* 43, no. 3 (1982): 598.
40 Lou Ann Patterson and John Defrain, 'Pronatalism in High School Family Studies Texts,' *Family Relations* 30, no. 2 (1981): 211.
41 Bayne and Kolers, 'Toward a Pluralist Account of Parenthood,' 235.
42 Madison Powers, 'Privacy and Genetics,' in *A Companion to Genethics*, eds Justine Burley and John Harris (Oxford: Blackwell, 2002), 368.
43 Sander-Staudt, 'Of Machine Born?' 116.
44 R.K. Sharma, et al., 'Sperm DNA Damage and Its Clinical Relevance in Assessing Reproductive Outcome,' *Asian Journal of Andrology* 6, no. 2 (2004): 140.
45 Debarun Majumdar, 'Choosing Childlessness: Intentions of Voluntary Childlessness in the United States,' *Michigan Sociological Review* 18 (2004): 111.
46 Ibid.
47 Linda R. Hirshman, *Get to Work: A Manifesto for Women of the World* (New York: Viking, 2006), 54.
48 Ibid., 24.
49 Ibid., 25.
50 Simonstein and Mashiach-Eizenberg, 'The Artificial Womb,' 88.
51 Hirshman, *Get to Work*, 58.
52 Oakley, 'Gender and Generation,' 29.
53 Nancy Folbre, *Who Pays for the Kids? Gender and the Structures of Constraint* (London: Routledge, 1994), 104.
54 Ruth Colker, *Pregnant Men: Practice, Theory, and the Law* (Bloomington: Indiana University Press, 1994), 159.

55 Australian Government Department of Human Services, 'Paid Parental Leave: Employee Eligibility,' 22 January 2015. Available at: http://www.humanservices.gov.au/ business/services/centrelink/paid-parental-leave-scheme-for-employers/employee-eligibility.
56 Robyn Rowland, 'Technology and Motherhood: Reproductive Choice Reconsidered,' *Signs* 12, no. 3 (1987): 524.
57 Carlo Bulletti, et al., 'The Artificial Womb,' *Annals of the New York Academy of Sciences* 1221 (2011): 127.
58 Bulletti, et al., 'The Artificial Womb,' 126.
59 Van Gelder, et al., 'Teratogenic Mechanisms of Medical Drugs,' 379.
60 Australian Bureau of Statistics, 'Causes of Death, Australia, 2010. Perinatal Deaths: Main Condition in Mother,' 14 March 2013. Available at: http://www.abs.gov.au/ausstats/abs@.nsf/Products/ C26E274706A50885CA2579C6000F73F1?opendocument.
61 M. Jean Heriot, 'Fetal Rights versus the Female Body: Contested Domains,' *Medical Anthropology Quarterly* 10, no. 2 (1996): 179.
62 Simonstein and Mashiach-Eizenberg, 'The Artificial Womb,' 88.
63 Gregory Pence, 'What's So Good about Natural Motherhood? (In Praise of Unnatural Gestation),' in *Ectogenesis: Artificial Womb Technology and the Future of Human Reproduction*, eds Scott Gelfand and John R. Shook (New York: Rodopi, 2006), 85.
64 Bayne and Kolers, 'Toward a Pluralist Account of Parenthood,' 224.
65 Firestone, *The Dialectic of Sex*, 197.
66 Parliament of Australia, *Prohibition of Human Cloning for Reproduction and the Regulation of Human Embryo Research Amendment Act 2006*, no. 172 (2006): 10–11.
67 Singer and Wells, *Making Babies*, 3; 6.
68 Ibid., 3. It is noted in the IVF literature of the early 1980s that two of the four most experienced clinical groups involved in this research were based in Melbourne, Australia (Clifford Grobstein, et al., 'External Human Fertilization: An Evaluation of Policy,' *Science* 222, no. 4620 [1983]: 127).
69 Gabor T. Kovacs, et al., 'Embryo Donation at an Australian University In-Vitro Fertilisation Clinic: Issues and Outcomes,' *Medical Journal of Australia* 178, no. 3 (2003): 127.
70 Amel Alghrani and Margaret Brazier, 'What Is It? Whose It? Re-Positioning the Fetus in the Context of Research?' *Cambridge Law Journal* 70, no. 1 (2011): 72.
71 C.M. Farquhar, et al., 'A Comparative Analysis of Assisted Reproductive Technology Cycles in Australia and New Zealand 2004–2007,' *Human Reproduction* 25, no. 9 (2010): 2281.

Background: The Story Thus Far

Abstract: *Kendal explores the history of humidicribs and in vitro fertilisation (IVF) as two of the existing technologies expected to aid the development of ectogenesis. Conservative and feminist reactions to reproductive biotechnologies are discussed, before addressing the representation of ectogenesis in popular media and its impact on public opinion. Finally, Kendal discusses recent advancements in ectogenesis research using animals and how these may inform human studies.*

Keywords: ARTs; assisted reproductive technologies; humidicribs; *in vitro* fertilisation; IVF; moral right; radical feminism

Kendal, Evie. *Equal Opportunity and the Case for State Sponsored Ectogenesis.* Basingstoke: Palgrave Macmillan, 2015. DOI: 10.1057/9781137549877.0004.

This section will provide a very brief summary of the current state of ectogenesis research. It is generally predicted that full ectogenesis, if it becomes possible in the future, will be achieved through the intersection of two existing technologies: IVF and artificial incubation. While legal restrictions currently prohibit the extension of the former beyond the first two weeks of embryonic development, practical constraints are implicated in the limitations of the latter, with the youngest recorded premature infant to survive being born at approximately 22 weeks gestation.[1]

A brief history of the development of humidicribs and *in vitro* fertilisation

Humidicribs have a somewhat sordid history. Following the Franco-Prussian War of 1870 French leaders became concerned about 'de-population,' particularly in the face of rising fertility rates among their chief rival, Germany.[2] According to Jeffery P. Baker, as a result of this, reducing infant mortality became a political issue, with the poor survival rates among infants born prematurely seen as 'a problem that robbed the nation of future workers and soldiers.'[3] The story continues that in 1878, Étienne Stéphane Tarnier visited the Jardin d'Acclimatation, was inspired by the chicken incubators on display designed by M. Odile Martin of the Paris Zoo, and requested that the zookeeper build a similar warming chamber large enough to fit one or two premature human infants. By 1880, the first warm air incubators were installed at the Paris Maternité Hospital and infant mortality began to decline.[4] Tarnier's warming system was not that dissimilar to the existing methods of maintaining the body temperature of premature infants, but he is still generally credited as providing the template upon which future incubators would be created.

Tarnier had a number of close colleagues, most notably including his successor, Pierre Budin, and Adolphe Pinard. Pinard was opposed to saving premature babies on eugenicist grounds, a position shared by many American obstetricians at the time in the absence of any concern regarding under-population.[5] However, Budin's advocacy kept incubators from being abandoned, and his association with 'physician-showman' Martin Couney was responsible for introducing the incubator-baby sideshows to America.[6] These shows, in which premature babies in incubators were displayed to the paying public as medical curiosities, were already popular in Europe, with Alexandre Lion, the designer of the more sophisticated (and substantially more expensive) Lion incubators of the 1890s displaying his invention at the Berlin Exposition of 1896 in an exhibition entitled the *Kinderbrutenstalt* ('child hatchery').[7] Couney's shows in America were initially very popular; however, several from the medical profession objected to the use of premature babies as sideshow attractions.[8] Couney would later become associated with Chicago-based prematurity expert, Julius Hess, who invented the Hess incubator in the early 1920s. Hess also developed a miniaturised version known as the 'Ambulance Box,' which was used to transport premature infants to hospitals for incubation.[9] Since then, the only major developments in humidicrib technology have involved improvements to ventilation systems and the creation of mobile intensive care units.[10] In the early 1990s, Nobuya Unno et al. hypothesised that placing premature infants in a fluid-based incubator would be preferable to existing air-based models; however, to date no such system has been developed.[11]

While humidicribs met with minimal opposition, save for a brief encounter with Social Darwinism in the early 20th century, on the other end of the spectrum IVF research has always been subject to rigorous ethical debate. The process of IVF involves harvesting a woman's ova by laparoscopy under anaesthesia, fertilising the selected ovum outside her body, then attempting to transfer the resultant embryo into her womb to establish a pregnancy. IVF treatment carries various risks for women, including ovarian hyperstimulation syndrome, infection, ectopic pregnancy, internal organ damage and hypertension.[12] In spite of these risks, by 2012 over four million babies worldwide had been born through this technology.[13]

Walter Heape achieved the first rabbit-to-rabbit embryo transfer in 1890.[14] Albert Brachet then made the first successful attempt to culture mammalian embryos *in vitro* in 1913. Gregory Pincus expanded on this research in the mid-1930s, achieving the first rabbit IVF pregnancy, which, unfortunately, afforded him such notoriety that he lost his position at Harvard University.[15] The public were so outraged at the possibility of IVF being used to create 'soulless' humans that Pincus was vilified in the *New York Times* and labelled 'Dr Frankenstein.'[16] Such antagonism later plagued Edwards and Steptoe, with public concern about IVF creating 'monstrous creatures' that were not fully human.[17] However, following Brown's uneventful birth and subsequent healthy development, IVF became widely tolerated.[18] Research regarding the physical and psychological development of children conceived through IVF has indicated no statistically significant differences when compared to naturally conceived controls, and infant attachment, parental competence and breastfeeding habits have also shown no difference.[19] Most of the ethical concerns now cited regarding IVF focus on the emotional and financial costs of the procedure, with the probability of a live birth still only estimated at 72 per cent for couples undergoing six IVF cycles.[20]

Conservative reactions to reproductive biotechnologies

Although humidicribs were generally welcomed, the arrival of the first IVF baby initiated moral panic among some conservative groups. The Roman Catholic Church already opposed all forms of interference with natural pregnancy, with Pope Pius IX decreeing in his 1869 bull *Apostolicae Sedis Moderationi* that abortion at any stage would automatically lead to excommunication.[21] Similarly, Pope Paul VI's 1968 encyclical *Humanae Vitae* only allowed periodic abstinence as a method of family planning, condemning all other contraceptive methods. In 1987, the Congregation for the Doctrine of the Faith released the instruction *Donum Vitae*, which officially banned members of the Church from using IVF, with its 2008 supplement *Dignitas Personae* also specifically prohibiting embryo transfer, donation and manipulation.[22] The rationale behind such bans was to remove third-party interference in procreation, an act that was considered sacred between spouses.[23] Another concern was the loss of embryonic life entailed in IVF, which was seen to violate the dignity of these human lives. The fact that 'embryonic wastage' is a natural part of all reproduction received little attention, except that the Church considered the *deliberate* destruction of embryos as equivalent to murder.[24] According to Catholic dogma, ensoulment and full personhood occur at the moment of conception and remain until the moment of natural death. For a short time the Church suggested that babies born of IVF would lack a soul; however, this idea was abandoned after it was shown that IVF children developed the same as naturally conceived children.[25] The Church has not revised its stance on IVF; however, research indicates that Catholics access IVF in similar numbers to non-Catholics,[26] and that many members of the clergy support these choices.[27]

In 2006, the Bioethics Committee of the Holy Synod of the Greek Orthodox Church affirmed similar objections to IVF as the Catholic Church.[28] Nevertheless, many other world religions have embraced this technology. Since 1980, Sunni Muslims have had fatwas permitting IVF for married couples using their own gametes, and since the late 1990s, Shiite Muslims have also been allowed to use donor gametes, as long as traditional Islamic codes for parenthood and inheritance are observed.[29] Facilitating this position is the fact that according to Islamic theology, ensoulment of the fetus does not occur at fertilisation but rather at a later point in development. This view is also held in Judaism, with IVF and embryo transfer officially supported by both the Ashkenazi and Sephardic chief rabbis of Israel and the Jewish majority.[30] Believers within Confucianism, Hinduism, Buddhism and many denominations within Protestant Christianity have also embraced IVF and other ARTs to varying degrees.[31] While it is difficult to speculate on how these religious groups would react to ectogenesis, it seems likely that many among the 'moral right' (both secular and religious) would oppose the advent of this technology because it challenges traditional notions of the family. Perhaps the only concessions that can be made are that ectogenesis would also save the lives of premature babies, provide an alternative to some abortions and eliminate the need for selective pregnancy reduction for IVF-induced multiple pregnancies. All of these satisfy various pronatalist, conservative agendas, but only by decentralising the goal of alleviating the burden of women's reproductive labour.[32] This aspect remains the focus for feminist arguments regarding this technology.

Feminist reactions to reproductive biotechnologies

Just as every religion has a different view on ARTs, so does every major feminist ideology. Members within the camp that consider technological interference in reproduction a threat to women's liberty include such pivotal figures as Robyn Rowland, Gena Corea, Andrea Dworkin and Barbara Katz Rothman.[33] Rowland voices concern that men will use ARTs to exert even more control over women's lives and that infertile women will be coerced into using new technologies that promote pronatalist agendas.[34] Several other feminist writers agree that men have a history of usurping the control of technology and imposing it unnecessarily on women's bodies.[35] Judith Lorber claims that for all society's focus on mothers, men are often the 'dominant partner in childbearing decisions,' suggesting the use of ARTs, particularly for male-factor infertility, represents a 'patriarchal bargain' many women must strike to maintain their relationships.[36] Corea goes so far as to suggest men will use ARTs to eliminate women entirely, or to create a breeder sub-class of women whose reproductive labour can be purchased in much the same way as sexual favours are exchanged in brothels.[37] On the other side of the spectrum are radical feminists like Firestone, Smajdor, Marge Piercy and Adrienne Rich, who argue that natural pregnancy and the social conditions surrounding it are the root cause of women's oppression; since pregnant women are cast as the 'archetypal altruists,' self-sacrifice is demanded of them in ways never expected of men.[38] It is within this group that ectogenesis finds its greatest allies.

According to Julien S. Murphy, some feminists are concerned that ectogenesis would undermine women's liberation efforts by appropriating one of women's more cherished contributions to society and resituating it within the realm of male-dominated science.[39] Others are disturbed by the potential loss of influence over childrearing that ectogenesis would entail for biological mothers, were they no longer so intimately connected with the process of gestation.[40] Yet there are also strong feminist arguments that suggest women's liberty would be increased if ectogenesis became available. Suzanne Dixon notes that patriarchal societies have often justified the subjugation of women on the grounds that patrilineal succession systems required a different standard of sexual behaviour be enforced for women compared with men, to ensure the 'purity of the male line of descent.'[41] That there was no other way to guarantee this in the past but to demand female chastity and severely punish instances of infidelity is one of the root causes of women's sexual oppression. When establishing a power base required exclusive rights to a woman's reproductive ability, social structures supporting the treatment of women as commodities (and limiting their reproductive liberty) were likely to abound. If, as Curtis claims, 'male dominance has been predicated on men's ability to control women's reproductive life,' it stands to reason that providing an alternative to men having to 'own' a woman's uterus for the purpose of procreation may in fact bring positive social change for many women.[42] I argue that the pronatalist construction of motherhood as being 'essential' to female identity is the symptom of a biological reality in which species propagation depends on encouraging women to be pregnant. As such, were ectogenesis available to alleviate some of this burden, women may be free to seek self-actualisation in more varied social arenas and on more equal footing with men.

Regarding the fear that ectogenesis would lead to a reduction in perceived value for *all* women in society, I argue this is not a justifiable reason to abandon the goal of ectogenesis, but rather indicates that other social changes need to accompany it, including an increased awareness of women's value in society above and beyond their reproductive capacity. Murphy argues that doomsday predictions, like Corea's, of a world in which technological reproduction has led to complete 'femicide' are unrealistic, as even within the non-ideal patriarchal societies most women live, women are still valued as companions and sexual partners.[43] While feminism strives to create a more equal world, in the short term this provides some reassurance that ectogenesis would not 'replace' the need for women in society.

Although my position in this book is predominantly a liberal feminist one, as it focuses on increasing reproductive options for women so they can engage more fully in public life, my argument will also appeal to some radical feminists. Marxist and socialist feminists are likely to support the egalitarian project underpinning my focus on equal opportunity, while conservative feminists may appreciate the potential for ectogenesis to serve particular conservative agendas, such as reducing the number of abortions procured for unwanted pregnancies. The major point separating my argument from a libertarian feminist view is the focus on providing state sponsored, rather than privately funded access to the technology. As it focuses on promoting individual choice, I argue the liberal feminist perspective has a particularly valuable contribution to make to the ART debate.

A brief history of ectogenesis in popular media

Popular media also contributes to the ectogenesis debate in society. J.B.S. Haldane first coined the term 'ectogenesis' in a lecture given to the Heretics Society of the University of Cambridge in 1923. A few years later the idea of artificial wombs became widely disseminated through Aldous Huxley's *Brave New World* (1932).[44] This text has had a profound impact on society, with references to Huxley's dystopian classic frequently appearing in bioethical and legal scholarship regarding ARTs.[45] Susan M. Squier notes that throughout the 1920s and 1930s ectogenesis featured prominently in various science fiction stories, including Haldane's *Daedalus, or Science and the Future* (1923), with the overwhelming majority of these tales depicting ectogenesis in a negative light.[46] Piercy's *Woman on the Edge of Time* (1976) is one of the few works that represents ectogenesis positively, but even here the loss of biological pregnancy is seen as a tragic sacrifice women had to make to achieve sexual equality.[47]

Jay Clayton claims that 'the horrors of science fiction haunt the bioethical imagination,' which goes some way to explaining the societal fears connected with emerging ARTs, like ectogenesis.[48] With horrific filmic representations of artificial wombs like those seen in *The Matrix* (1999) and *The Island* (2005) permeating popular culture, ectogenesis has acquired somewhat of an image problem. It is not just in dystopian fiction that this anti-technological bias plays out either. Society's rigid perception of women's role in procreation is perhaps most evident when exploring utopian imaginings of the future in which brain surgery may be performed with the flick of a switch, but women are still screaming through the extreme pain of labour and childbirth (this occurs, e.g., in various *Star Trek* episodes from the 1990s through to the most recent film series). That achieving immortality through biotechnology is often portrayed as being more likely than escaping the demands of physical gestation illustrates an unwillingness to challenge the norm that women should suffer as a result of their reproductive capacity.

The scientific possibility of ectogenesis

Before discussing in more detail the potential benefits of ectogenesis for women, it is first necessary to consider whether or not the technology is likely to become a reality. Scientists have been attempting to create an artificial uterus since the 1950s; however, significant advances have been made only in recent years.[49] Ectogenesis is already practiced in animal research, including a recent species preservation project in Australia aimed at preventing the extinction of the grey nurse shark. Successful artificial incubation was achieved in the prototype acrylic tank using late stage embryos of the smaller wobbegong shark, resulting in viable offspring that did not differ substantially from naturally gestated controls.[50] The scientist responsible for the project, Dr Nick Otway, has gone on record saying the device could be adapted for use in other species;

however, 'he has no plans to make artificial human uteruses... because of the ethical issues.'[51] Over the last decade human uteri have also been successfully extracted and functionally maintained for limited periods of time.[52] Survival times for embryos sustained in artificial or transplanted uteri have typically been very low though, ranging from hours to weeks.[53] Thus far the single greatest obstacle to full human ectogenesis has been the need to produce an artificial placenta, capable of supplying nutrients and blood to the developing fetus. While it has already been established that a transplanted human placenta is capable of functioning when attached to an artificial uterus, further research is required before the process of fetal development can be sustained entirely *ex vivo*.[54]

Scientists for the New York Academy of Science released a statement in 2011 claiming that *partial* ectogenesis, which they defined as artificial development of a fetus between 14 and 35 weeks gestation, was 'within reach given our current knowledge and existing tools.'[55] It is also believed that advances in fetal medicine make it likely surgeons will soon be able to surgically extract a partially formed fetus from its mother's womb and transfer it into an artificial incubator with no ill effects.[56] When it comes to developing *full* human ectogenesis, researchers generally agree that there are three distinct technical requirements that would need to be achieved in order for an artificial womb to serve as a complete substitute for *in utero* gestation. The first is the need to develop a three-dimensional 'shell' to act in place of the woman's uterus and provide a site for embryo implantation.[57] Experimental chambers have already been constructed from explanted uterine tissue, synthetic materials and human endometrial cells grown on biodegradable scaffolding.[58] While animal studies have shown promising results, experiments using human embryos in these artificial wombs have been impeded by what some scientists refer to as the 'obvious ethical concerns' regarding the length of time an embryo should be sustained in such an environment.[59] Primary amongst these concerns seems to be the moral status of the ectogenetic embryo or fetus and the complications regarding its parentage. The second requirement for full ectogenesis would be the development of a suitable replacement for amniotic fluid. This aspect is generally considered the least problematic as artificial amniotic fluid has already been created and used successfully in animal experiments.[60] As noted earlier, however, the third requirement constitutes the major obstacle to ectogenesis research so far – the need to create an artificial placenta that could function as a regulatory system supplying the appropriate levels of blood, nutrients, hormones and oxygen to the developing fetus. While providing the first three is achievable given existing methods of intravenous feeding and amniotic fluid exchange, oxygenation remains a significant issue, both in theoretical ectogenesis research and practical neonatal care.

Currently, the survival of a baby born preterm depends on the maturity of its lungs, and whether an oxygen pump or extracorporeal membrane oxygenator (ECMO)[61] can provide sufficient air to keep it alive.[62] This is because, despite the fact the fetus has its own independent blood supply *in utero*, it is entirely dependent on maternal oxygen.[63] Until recently, ECMO has shown limited success in treating premature babies, with the size of components making it unsuitable for very small infants.[64] There have been significant breakthroughs made in this area since 2011, with Jutta Arens et al. announcing the development of a miniaturised oxygenator that will allow, they believe, for a system of extracorporeal gas exchange similar to that which naturally occurs between the fetus and placenta.[65] While the immediate benefits of this technology will apply to premature neonates, there is also the potential for this system to be adapted for use within an ectogenesis chamber. The authors describe their new development, dubbed the NeonatOx system, as 'one step on the way to the clinical application of the artificial placenta.'[66] There is also the possibility, given our knowledge of embryo implantation in ectopic pregnancy, that if the artificial 'shell' used to house the ectogenetic fetus were perfused with sufficient blood and other nutrients, the fetus may be able to create an organic placenta itself.[67] While an ectopic pregnancy is generally considered an obstetric emergency requiring termination, in the rare cases where one has gone to term, the lack of sufficient amniotic fluid protecting the fetus has been believed to be the primary cause of any malformation, and not any incompetence of its atypically placed placenta.[68] I argue another potential benefit of ectogenesis would be that a wanted ectopic pregnancy could be continued artificially after removal, which is particularly relevant when considering that the extraction can have a negative impact on the woman's future fertility, with fallopian tube resection often being a necessary countermeasure to prevent future complications.

Once the three technical requirements outlined are satisfied, what remains to be considered is how the genetic material needed to create the ectogenetic offspring would be obtained. The least contentious method, and the one most likely to advance ectogenesis research in the earlier stages, is the transfer of babies born increasingly premature into artificial incubation chambers. This method could be easily justified on the grounds of providing some chance at life for these infants. The second option is fetal transfer, the removal of a partially formed fetus from its mother's womb in order to continue its gestation artificially. Again, the development of this process is likely to stem from medical emergencies, although it could also originate from attempts to 'rescue' aborted fetuses. There are many bioethicists who believe imposing the risks of unproven ectogenesis technology could be defensible at first only if limited to cases where the fetus is otherwise destined for abortion.[69] The third option would be embryo transfer following either *in vitro* or *in vivo* fertilisation. IVF initiation would carry with it many of the usual health-related concerns regarding this technology. Using this method, a woman (whether the intended mother or a donor) would still need to submit to the physical risks associated with hormonal overstimulation and invasive surgical procedures, but would not need to undergo embryo transfer or pregnancy as the embryo would be transferred to the ectogenesis chamber for development. As embryo transfer is the least successful element of standard IVF and causes significant distress for women whose embryos do not implant, this method provides an alternative to having the highly sought after embryo die within the hopeful mother.[70] The problems associated with IVF and multiple pregnancies could also be prevented, by transferring embryos one at a time to the artificial womb until one successfully implants. Without the need to prepare the woman's uterus for implantation, such attempts could be repeated immediately and with no negative consequences. Assuming the risks associated with IVF in its current form are ethically acceptable, there is no compelling reason to prohibit IVF initiation of ectogenesis, where these risks are arguably diminished. Indeed, consistency would demand that so long as the first use of IVF is permissible, there could be no grounds to prohibit the second on the basis of potential risk to the mother, presupposing autonomous and informed consent in both cases. While using ectogenesis following IVF cannot eliminate all the physical risks to women choosing to reproduce through this means, the physical burden of IVF alone is certainly less than the physical burden of IVF *and* pregnancy. For women who would need to rely on IVF to achieve pregnancy anyway, the option of artificial gestation still represents a significant risk reduction. This is particularly compelling when considering that physical pregnancy after IVF is

Background: The Story Thus Far   37

known to involve higher risks of maternal complications, including preeclampsia, placenta abruption and preterm delivery.[71]

To completely avoid the risks involved with IVF, some women could reap the benefits of egg donation instead. Singer and Wells note that currently many IVF patients are happy to donate surplus eggs to other infertile women, precisely because they understand the 'misery infertility can cause'.[72] Furthermore, they note women undergoing tubal sterilisation could choose to donate eggs without any additional surgical procedures. Some IVF clinics already have 'egg sharing' programmes in effect where a woman who lacks healthy ova can subsidise the costs of another woman's treatment in exchange for some of her eggs.[73] However, those women who want their baby born through ectogenesis to be their own genetic offspring would still need to provide the eggs themselves. An alternative to IVF initiation for these women could be embryo transfer following *in vivo* fertilisation, which would be less invasive than extracting and transferring a partially developed fetus. In this method fertilisation occurs either via sexual intercourse or artificial insemination, and then the embryo is recovered and transferred to the desired implantation site.[74] In the case of embryo donation this would be the intended mother's uterus, for ectogenesis, the artificial womb. In 2014, K. Pagidas et al. showed a uterine lavage catheter could be used to recover fertilised mouse embryos from simulated silicon uteri with 96.7 per cent of recovered blastocysts deemed viable one day later. The authors also reported no tissue abrasion or punctures to extirpated human uteri following the same lavage technique, ultimately concluding that this device is ready for use in human trials to allow for the possibility of genetically testing *in vivo* derived embryos before implantation.[75] Although this technique of obtaining embryos has far fewer risks to the mother than IVF, timing has to be closely monitored, as there is the potential for the embryo to implant before it can be recovered. This currently poses a significant problem for embryo donation, but not necessarily for ectogenesis, as the fetus could simply be transported across as soon as it was surgically safe to do so. The potential for fetal transport is also important for women who do not discover they are pregnant until after implantation has occurred.

Given the tools currently available in neonatology and the ever-expanding field of ARTs, I argue that it is likely ectogenesis could become a reality provided there's the necessary legislative and funding support. Although estimated time frames vary from 12 months to 30 years, most scientists working on developing artificial means of gestation agree that full human ectogenesis is a real scientific possibility.[76] Even if the research possibilities outlined in the previous pages turn out to be unsuccessful, alternatives will develop in their place. One question that is often raised when debating the ethics of ectogenesis research is whether the technology would provide a benefit substantial enough to justify the expense of developing it. Throughout this book it will be my contention that the ideals of equal opportunity can be used to argue in favour of continued research in this area. While it is undoubtedly necessary to improve maternal healthcare and limit the negative outcomes associated with pregnancy and childbirth, I argue these alone would not guarantee equal opportunity for women, and therefore ectogenesis would still have a unique role to play in providing women with reproductive alternatives. Importantly, it is not necessary to defend the naively optimistic view that ectogenesis technology would solve all, or even the majority, of gender-based inequalities related to pregnancy, childbirth and parenthood identified in this book – for the purposes of justifying the research, it is sufficient that it redress some.

Notes

1   Amillia Taylor was conceived through IVF and thus her gestational age was known precisely. Born at 21 weeks and 6 days gestation she was artificially incubated until hospital discharge (S.R. Doherty, 'Could We Care for Amillia in Rural Australia?' *Rural and Remote Health* 7, no. 4 [2007]: 768.).
2   Jeffery P. Baker, 'The Incubator Controversy: Pediatricians and the Origins of Premature Infant Technology in the United States, 1890 to 1910,' *Pediatrics* 87, no. 5 (1991): 654.
3   Jeffery P. Baker, 'The Incubator and the Medical Discovery of the Premature Infant,' *Journal of Perinatology* 5 (2000): 323.
4   William A. Silverman, 'Incubator-Baby Side Shows,' *Pediatrics* 64, no. 2 (1979): 128.
5   Baker, 'The Incubator Controversy,' 655.
6   Baker, 'The Incubator,' 323.
7   Ibid., 324.
8   Silverman, 'Incubator-Baby Side Shows,' 130.
9   Tonse N.K. Ragu, 'An Extant Hess Incubator on Display,' [letter to the editor] *Pediatrics* 107, no. 4 (2001): 805.
10 Richard J. Martin, et al. *Fanaroff and Martin's Neonatal-Perinatal Medicine: Diseases of the Fetus and Infant*, 10th ed. (St Louis: Elsevier, 2015).

11 Nobuya Unno, et al., 'Development of an Artificial Placenta: Survival of Isolated Goat Fetuses for Three Weeks with Umbilical Arteriovenous Extracorporeal Membrane Oxygenation,' *Artificial Organs* 17, no. 12 (1993): 1002.
12 V. Beral, et al., 'Outcome of Pregnancies Resulting from Assisted Conception,' *British Medical Bulletin* 46, no. 3 (1990): 757; 759.
13 John D. Biggers, 'IVF and Embryo Transfer: Historical Origin and Development,' *Reproductive Biomedicine Online* 25, no. 2 (2012): 118.
14 M.C. Chang, 'My Work on the Transplantation of Mammalian Eggs,' *Theriogenology* 19, no. 3 (1983): 293.
15 Biggers, 'IVF and Embryo Transfer,' 119–20.
16 Anil K. Dubey, *Infertility: Diagnosis, Management, & IVF* (New Delhi: Jaypee Brothers Medical Publishers, 2012), 17.
17 Giuseppe Benagiano, et al., 'Robert G Edwards and the Roman Catholic Church,' *Reproductive Biomedicine Online* 22 (2011): 670.
18 Chris Mason, 'Making People: Today's Wariness of Reproductive Technologies Stems from Myths, Legends and Hollywood,' *Nature* 471, no. 7338 (2011): 299.
19 Hilde Colpin, 'Parenting and Psychosocial Development of IVF Children: Review of the Research Literature,' *Developmental Review* 22, no. 4 (2002): 644–73; Catherine A. McMahon, et al., 'Psychosocial Adjustment and the Quality of the Mother-Child Relationship at Four Months Postpartum after Conception by In Vitro Fertilization,' *Fertility and Sterility* 68, no. 3 (1997): 492–500.
20 Nicolás Garrido, et al., 'Cumulative Live-Birth Rates Per Total Number of Embryos Needed to Reach Newborn in Consecutive In Vitro Fertilization (IVF) Cycles: A New Approach to Measuring the Likelihood of IVF Success,' *Fertility and Sterility* 96, no. 1 (2011): 40.
21 Norman M. Ford, 'A Catholic Ethical Approach to Human Reproductive Technology,' *Reproductive BioMedicine Online* 17, sup. 3 (2008): 40.
22 Katha Pollitt, *Pro: Reclaiming Abortion Rights* (New York: Picador, 2014), 70–89.
23 Paul T. Schotsmans, 'In Vitro Fertilisation: The Ethics of Illicitness? A Personalist Catholic Approach,' *European Journal of Obstetrics & Gynecology and Reproductive Biology* 81 (1998): 240.
24 D. Gareth Jones, *Brave New People: Ethical Issues at the Commencement of Life* (Leicester: Inter-Varsity Press, 1984), 121.
25 Lawrence Krauss, 'Science the Catholic Church Can't Ignore,' *New Scientist*, 7 February 2009, 25.
26 Paul Lauritzen, 'Catholics & IVF: The Next Big Battleground?' *Commonweal*, 12 August 2005, 10.
27 Pierre Mallia, 'Problems Faced with Legislating for IVF Technology in a Roman Catholic Country,' *Medicine, Health Care and Philosophy* 13 (2010): 81.

28 Benagiano, et al., 'Robert G Edwards and the Roman Catholic Church,' 667.
29 Michael Brooks, 'Faith in Denial,' *New Scientist*, 26 July 2008, 18; Marcia C. Inhorn, 'Making Muslim Babies: Ivf and Gamete Donation in Sunni versus Shi'a Islam,' *Culture, Medicine and Psychiatry* 30, no. 4 (2006): 427.
30 Benagiano, et al., 'Robert G Edwards and the Roman Catholic Church,' 668; J.G. Schenker, '*In Vitro* Fertilization and Embryo Transfer: Jewish Ethical and Legal Aspects,' in *In Vitro Fertilization, Embryo Transfer and Early Pregnancy*, eds R.F. Harrison, J. Bonnar and W. Thompson (Lancaster: MTP Press Limited, 1984), 181–1. As in the case of Muslim couples, the use of donor gametes causes some issues in relation to kinship ties.
31 Elizabeth F.S. Roberts, 'God's Laboratory: Religious Rationalities and Modernity in Ecuadorian In Vitro Fertilization,' *Culture, Medicine and Psychiatry* 30, no. 4 (2006): 507.
32 Pamela Schaeffer, 'In Vitro Fertilization Widely Used,' *National Catholic Reporter*, 15 October 1999, 13–16.
33 Rosemary Tong, *Feminist Thought: A Comprehensive Introduction* (Boulder, Colorado, USA: Westview Press, 1989), 81.
34 Robyn Rowland, 'A Child at Any Price? An Overview of Issues in the Use of the New Reproductive Technologies, and the Threat to Women,' *Women's Studies International Forum* 8, no. 6 (1985): 540.
35 Jan Zimmerman, 'Technology and the Future of Women: Haven't We Met Somewhere Before?' *Women's Studies International Forum* 4, no. 3 (1981): 364; Rita Arditti, 'Reducing Women to Matter,' *Women's Studies International Forum* 8, no. 6 (1985): 577.
36 Judith Lorber, 'Choice, Gift, or Patriarchal Bargain? Women's Consent to *In Vitro* Fertilization in Male Infertility,' in *Feminist Perspectives in Medical Ethics*, eds Helen Bequaert Holmes and Laura M. Purdy (Bloomington: Indiana University Press, 1992), 177.
37 Genoveffa Corea, 'How the New Reproductive Technologies Could be Used to Apply the Brothel Model of Social Control over Women,' *Women's Studies International Forum* 8, no. 4 (1985): 299.
38 Janice G. Raymond, 'Reproductive Gifts and Gift Giving: The Altruistic Woman,' in *Life Choices: A Hastings Center Introduction to Bioethics*, 2nd ed., eds Joseph H. Howell and William Frederick Sale (Washington D.C.: Georgetown University Press, 2000), 398.
39 Murphy, 'Is Pregnancy Necessary?' 67.
40 Rowland, 'Technology and Motherhood,' 515.
41 Suzanne Dixon, 'Conclusion – The Enduring Theme,' in *Stereotypes of Women in Power: Historical Perspective and Revisionist Views*, eds Barbara Garlick, Suzanne Dixon and Pauline Allen (New York: Greenwood Press, 1992), 213.
42 Curtis, 'Hannah Arendt,' 162.
43 Murphy, 'Is Pregnancy Necessary?' 76.

44 Biggers, 'IVF and Embryo Transfer,' 119.
45 Kieran Tranter, 'The Speculative Jurisdiction: The Science Fictionality of Law and Technology,' *Griffith Law Review* 20, no. 4 (2011): 827.
46 Susan M. Squier, *Babies in Bottles: Twentieth-Century Visions of Reproductive Technology* (New Brunswick: Rutgers University Press, 1994), 67.
47 Tong, *Feminist Thought*, 76.
48 Jay Clayton, 'The Ridicule of Time: Science Fiction, Bioethics, and the Posthuman,' *American Literary History* 25, no. 2 (2013): 318.
49 Megan-Jane Johnstone, 'Ethics and Ectogenesis,' *Australian Nursing Journal* 17, no. 11 (2010): 33.
50 Nick Otway and Megan Ellis, 'Construction and Test of an Artificial Uterus for Ex Situ Development of Shark Embryos,' *Zoo Biology* 31, no. 2 (2012): 197–205.
51 Corey Binns, 'The Shark Factory: An Artificial Uterus Gives an Endangered Species a Shot at Survival,' *Popular Science* 275, no. 1 (2009): 34.
52 Johnstone, 'Ethics and Ectogenesis,' 33.
53 Ibid.
54 Bulletti, et al., 'The Artificial Womb,' 125.
55 Ibid., 127.
56 Jessica H. Schultz, 'Development of Ectogenesis: How Will Artificial Wombs Affect the Legal Status of a Fetus or Embryo?' *Chicago-Kent Law Review* 84 (2010): 880.
57 P. Chavatte-Palmer, et al., 'Une Reproduction Sans Utérus? État des Lieux de L'ectogenèse' [Reproduction without a Uterus? State of the Art of Ectogenesis], *Gynécologie Obstétrique and Fertilité* 40, no. 11 (2012): 695.
58 Schultz, 'Development of Ectogenesis,' 879. Leading researchers include Hung-Ching Liu of Cornell University and Yoshinori Kuwabara of Juntendo University (Bulletti, et al., 'The Artificial Womb,' 125).
59 Chavatte-Palmer, 'Une Reproduction Sans Utérus?' 695.
60 Schultz, 'Development of Ectogenesis,' 880.
61 ECMO is used to provide oxygen to premature infants and adults suffering cardiac and respiratory failure. Through a variety of techniques blood is extracted, oxygenated outside the body and then infused back into the patient.
62 Jutta Arens, et al., 'NeonatOx: A Pumpless Extracorporeal Lung Support for Premature Neonates,' *Artificial Organs* 35, no. 11 (2011): 997.
63 Bayne and Kolers, 'Toward a Pluralist Account of Parenthood,' 230.
64 Chavatte-Palmer, 'Une Reproduction Sans Utérus?' 695.
65 Arens, et al., 'NeonatOx,' 998.
66 Ibid., 1000.
67 Robert Sparrow, 'Is It "Every Man's Right to Have Babies If He Wants Them?" Male Pregnancy and the Limits of Reproductive Liberty,' *Kennedy Institute of Ethics Journal* 18, no. 3 (2008): 276.

68 Otto Bruckschwaiger, 'Four Cases of Advanced and Full-Term Ectopic Pregnancy,' *Canadian Medical Association Journal* 76, no. 9 (1957): 761; Alexandria J. Hill, et al., 'A True Cornual (Interstitial) Pregnancy Resulting in a Viable Fetus,' *Obstetrics and Gynecology* 121, no. 2 (2013): 428.
69 Alghrani and Brazier, 'What Is It?' 78.
70 Singer and Wells, *Making Babies*, 83.
71 Ioanna Tsoumpou, et al., 'Failed IVF Cycles and the Risk of Subsequent Preeclampsia or Fetal Growth Restriction: A Case-Control Exploratory Study,' *Fertility and Sterility* 95, no. 3 (2011): 977.
72 Singer and Wells, *Making Babies*, 60.
73 Celia Roberts and Karen Throsby, 'Paid to Share: IVF Patients, Eggs and Stem Cell Research,' *Social Science & Medicine* 66 (2008): 159–69.
74 The human embryo recovery technique was pioneered by John Buster in 1984 and was adapted from the bovine uterine lavage used by commercial cattle breeders to flush out infectious material from within the reproductive tract (Judith Lasker, *In Search of Parenthood: Coping with Infertility and High-Tech Conception* [Philadelphia: Temple University Press, 1994], 94).
75 K. Pagidas, et al. 'Previvo Uterine Lavage Catheter: A Novel Device for the Recovery of In Vivo Derived Human Embryos by Non-Surgical Lavage,' *Fertility and Sterility* 102, no. 3, sup. (2014): e33.
76 Eric Steiger, 'Not of Woman Born: How Ectogenesis Will Change the Way We View Viability, Birth, and the Status of the Unborn,' *Journal of Law and Health* 23, no. 2 (2010): 144.

1
Promoting Equal Opportunity through Ectogenesis

Abstract: *Kendal addresses various threats to equal opportunity arising from women's reproductive capacity, including inequalities between men and women, women and other women (including the fertile and infertile) and between heterosexual and homosexual individuals. She posits that ectogenesis represents a necessary addition to the current suite of assisted reproductive technologies (ARTs) available in fertility care.*

Keywords: ARTs; assisted reproductive technologies; gender discrimination; infertility; sexual equality

Kendal, Evie. *Equal Opportunity and the Case for State Sponsored Ectogenesis*. Basingstoke: Palgrave Macmillan, 2015. DOI: 10.1057/9781137549877.0005.

As discussed in the Introduction, there are various physical, social and economic burdens associated with pregnancy and childbirth that can disadvantage women, thus highlighting the need to explore alternative methods of procreation. From an equal opportunity standpoint it must be recalled that while society as a whole benefits from the reproductive labour of women, the risks associated with pregnancy and childbirth are borne by the individual woman alone. This unfair distribution of risks and benefits leads Smajdor to claim that there has been 'a conceptual failure in medical and social and ethical terms to address the pathological nature of gestation and childbirth and to tackle the health problems it poses from a justice perspective.'[1] Although there are certain social and financial restrictions that impact both men and women in their reproductive endeavours, these are often distributed unequally to women's disadvantage. In this chapter I will explore various threats to equal opportunity arising from women's reproductive capacity, including inequalities between men and women, women and other women (including the fertile and infertile) and between heterosexual and homosexual individuals. I will also posit that ectogenesis could redress several of the biologically based inequalities related to gestation, childbirth and parenthood.

Promoting equality with men

In her article 'In Defense of Ectogenesis,' Smajdor notes: 'Pregnancy is a condition that causes pain and suffering, and that only affects women. The fact that men do not have to go through pregnancy to have a genetically related child, whereas women do, is a natural inequality.'[2] Many would argue that this is simply natural biology and that it is therefore counterintuitive to consider it a form of sex-based inequality. I maintain that this perspective fails to recognise that modern medicine is premised on *fighting against* our nature and biological shortcomings, particularly our 'natural' susceptibility to disease and the inevitability of death. While ageing and growing infirm may also be considered 'natural' biological events, the goal of modern geriatric medicine has long been to increase life expectancy and reduce the impact and occurrence of age-related chronic conditions, such as incontinence, sensory loss and neurodegeneration.[3] There is no compelling reason to relegate pregnancy and childbirth to the realm of inescapable biological difference, while medical science still strives to dominate other aspects of human biology. As Firestone rightly asserts, 'the "natural" is not necessarily a "human" value. Humanity has begun to outgrow Nature: we can no longer justify the maintenance of a discriminatory sex class system on the grounds of its origin in Nature.'[4] That women must submit to greater risks than men in order to achieve the same goal of having biologically related children is an example of sex inequality, regardless of the fact this inequality originates in Nature. I propose that while no one can be directly blamed for such an inequality, deliberate attempts to forestall or prevent the development of equalising alternatives *can* be considered ethically questionable. As Helen H. Lambert argues, 'If a particular sex difference is incompatible with important aspects of social equality, we should argue for compensatory measures, independent of biological causation.'[5] Furthermore, were ectogenesis to fail to become a viable alternative to pregnancy, I argue there would still be just cause to consider pregnancy and childbirth as unfairly disadvantaging women, even when no other avenue for perpetuating the species were available. In this case, other 'compensatory measures' would need to be taken to promote equality.

It is not necessary to attempt to classify pregnancy as a 'disease' in order to make its potential eradication a *medical* issue, as this would be a move that would likely alienate many feminists and fail to achieve the goal of improving conditions for pregnant women. The parasitic nature of the fetus and the physical burden of gestation on the individual woman are sufficient grounds to justify research into less physiologically demanding means of procreating, without the need to categorise pregnancy *per se* as a state of ill-health. Rather than denigrating pregnancy, I argue there is a need to address the health inequalities posed by reproductive difference, including pregnancy-related illness and injury. This is especially relevant when considering how large a portion of the human population are susceptible to these particular afflictions, and that alternative technologies, such as ectogenesis, may soon be available that could alter our fundamental assumptions regarding human reproductive destiny. I agree with Smajdor's assertion that such a radical re-evaluation of biological roles is currently hindered by the steadfast cultural belief in the 'necessity of women for gestating and nurturing society's children,' which she claims is 'so entrenched in our consciousness that we fail to recognize that we could change the situation.'[6] I believe this lack of imagination serves as an obstacle to ectogenesis research and impacts how pregnancy and childbirth are viewed in society.

When considering the hostility that may attend the advent of ectogenesis technology, Gregory Pence notes that all reproductive technologies have met with some disapproval at the outset, with the doctors responsible for performing the first artificial insemination even being labelled as 'perverts' by some social groups. He also notes that providing pain relief in childbirth was initially condemned by many religious institutions, including the Catholic Church, in what may seem to be a deliberate attempt to perpetuate women's suffering as a result of their biological difference.[7] Even now medical practitioners may be forced to defend their support of pain relief in childbirth amidst claims that such interventions defy the will of the God, or do not constitute a medical necessity for insurance purposes.[8] While many advances in reproductive technology meet with fierce social disapprobation, Firestone notes that those that reaffirm traditional, male-dominated social structures are often met with less opposition than those intended to ease women's burdens in procreation. For example, artificial insemination using sperm from a woman's husband has traditionally been considered more acceptable than artificial insemination by donor, particularly if the recipient does not conform to the conventional nuclear family structure (often determined by the absence of a 'father figure' for the resultant child, as in the case of lesbian couples or single women). For Firestone, the development of full human ectogenesis is a necessary step in securing gender equality, through alleviating women's suffering in pregnancy and childbirth and challenging ideals of 'male supremacy' and the sanctity of 'the family.'[9]

Although some of the burdens of pregnancy and childbirth have already been discussed in the Introduction, there are others that are also relevant to this debate. For example, in order to have their own biological children, women must be willing to take risks with their mental health that are not trivial. The emotional strain of pregnancy is well documented and among researchers there is increasing interest in the aetiology of postpartum depression, and the more severe postpartum psychosis, with some studies indicating a sevenfold increase in the number of psychiatric admissions occurring in the 30 days following childbirth.[10] Although the causes of both these conditions are unclear, there is reason to believe there is a direct link between the metabolic and hormonal changes caused during pregnancy and childbirth, and the onset of psychiatric symptoms.[11] Even among women who will not fall prey to these conditions, pregnancy and childbirth are known to cause significant emotional and social upheaval, in addition to the risk of short- and long-term physical damage. It should not be dismissed that childbirth is often used as a measure to describe the worst pain imaginable.[12] That this pain is considered women's 'fate' shows a shocking lack of concern for women's welfare. When discussing the frequency with which women are denied effective pain relief in labour, the American Congress of Obstetrics and Gynecology notes: 'There is no other circumstance where it is considered acceptable for an individual to experience untreated severe pain, amenable to safe intervention, while under a physician's care.'[13] The psychological trauma associated with the physical experience of childbirth, in addition to the hormonal changes brought on during this time, places strain on gestational mothers that is over and above that already accompanying the transition to parenthood. That men can become biological parents without submitting to the physical and psychological risks directly associated with pregnancy and childbirth while women are required to make these sacrifices to achieve the same goal is *prima facie* injustice.

The unequal 'cost of parenthood' exacted from men and women may alone justify the need to develop technology for *ex vivo* gestation.[14] Furthermore, pregnancy and childbirth involve *unique* threats to the mother's bodily integrity and personal privacy that do not affect the father.[15] When considering that the overwhelming majority of women in industrialised countries give birth in hospitals, Oakley notes: 'The fact that childbirth equates with patienthood often escapes unnoticed, but the role of the patient is a very particular one in our society, and it implies an ideal type of behaviour (passivity).'[16] There is no comparable institutionalisation that afflicts prospective fathers, which is particularly relevant when considering the potential occupational disadvantages affecting childbearing women.

Colker notes that despite the disproportionate physical risk and inconvenience impacting women in reproductive endeavours, gestational labour is often disregarded in court cases judging contested parentage. When reviewing a large number of legal cases in the United States, she found sperm donors have often been 'considered to have equivalent or superior claims to women who are gestational or biological mothers.'[17] Thus, when women have had to undergo pregnancy and childbirth to produce genetic offspring, men have been granted equal rights to those offspring for the mere donation of their sperm. Singer and Wells note that in IVF cases, such genetic input is 'easily obtained by masturbation,' whereas ova harvesting is a much more invasive procedure.[18] Nevertheless, in numerous cases where prospective parents have disagreed about the use or disposal of surplus embryos after IVF, equal genetic contribution has been the only legal consideration.[19] However, even this approach to legal parenthood has often been disregarded in order to privilege men's claims to genetic offspring over women's. Colker notes that in surrogacy cases where the gestational mother has petitioned for parental rights, regardless of whether she also donated the egg used to create the child, the US court system has almost always granted custody to the commissioning parents, often on the basis of the male partner's genetic relationship to the child.[20] As such, it appears the increase in the physical burden of childbearing for women, including ova harvesting for IVF pregnancies, does not carry an increase in the legal rights toward the resultant offspring. So-called sweat equity arguments aimed at prioritising parental rights on the basis of gestational labour are attempting to redress this; however, ectogenesis could go a lot further toward equalising the demands of procreation for men and women.

In addition to the social disadvantages associated with hospitalisation, there are other ways in which women's autonomy and bodily integrity are under threat in reproduction. These include the ever-increasing number of sanctions for activities believed to constitute 'fetal abuse,' such as taking drugs while pregnant. According to M. Jean Heriot, although these sanctions are ostensibly aimed at protecting fetal well-being, they are unfairly weighted to women's disadvantage. She notes that in many cases men are absolved from any responsibility for avoidable fetal damage 'as long as the neglect or abuse could be attributed to actions taken by the woman during pregnancy.'[21] Furthermore, she claims laws that punish women for 'negative outcomes' in pregnancy, such as the creation of a drug-addicted newborn, are 'based upon a rationale that makes the outcome of a pregnancy *uniquely the responsibility of the individual woman*.'[22] This 'sexist bias' is particularly of concern when one considers the dramatic increase in domestic violence that correlates with pregnancy, also known to harm the fetus but not specifically covered under 'fetal abuse.'[23] Although under-reporting is likely, instances of physical violence are known to occur against approximately 10–20 per cent of pregnant women in industrialised countries, indicating another area of physical risk associated with women's reproductive capacity.[24] According to Sander-Staudt, domestic violence statistics indicate women are more likely to be beaten when pregnant than at any other time and that 'blows inflicted against pregnant women are more likely to be aimed at their abdomens.'[25] She notes ectogenesis may help reduce this specifically pregnancy-related risk to women's personal safety, by creating 'less opportunity for others to view women's pregnant bodies as personal property, or a justifiable target for physical violence.'[26] While it would be highly inappropriate to suggest the solution to domestic violence is avoiding pregnancy, it is still the case that many pregnant women are subjected to domestic abuse *only* when their male partners are aware they are pregnant.[27] This supplies further evidence of women's vulnerability at this time; however, it does not negate the need for better protection for *all* women from domestic violence, whether they are pregnant or not. In *Pregnant Women Violent Men*, Sheila C. Hunt and Ann M. Martin claim that pregnancy is itself a trigger for domestic violence, with some women experiencing their first encounter with abuse following the announcement of a pregnancy. Unfortunately, following this initial trigger, many of these women will be the victims of assault for the remainder of their intimate relationship with this partner, often now with the additional complication of having a child involved.[28] As both a women's health and a legal issue, domestic violence deserves more attention, particularly regarding its complex relationship with pregnancy. The fact that women's behaviour while pregnant is so tightly regulated, while men's abuse of pregnant women often goes unpunished, is yet another example of how pregnancy is used to subjugate the interests of women and keep them in a state of vulnerability.

That men can reproduce without compromising their personal safety or being held responsible for inflicting 'fetal abuse' is a prime example of the inequality between men and women in reproduction.[29] When added to the unequal distribution of the physical, social and financial burdens associated with pregnancy and childbirth, the claim that reproductive difference constitutes a fundamental inequality between the sexes becomes especially convincing. While women must submit to these risks and suffer the consequences of any adverse complications in order to beget biological offspring, men can father children while retaining complete bodily integrity, privacy and freedom of movement. Singer and Wells therefore declare that ectogenesis 'can be supported on the ground that it would make a fundamental contribution toward sexual equality.'[30]

Given these concerns, it is perhaps not surprising that some feminists view artificial reproduction as a prerequisite for equality between the sexes. In her utopian vision of the future, Firestone foresees that 'the reproduction of the species by one sex for the benefit of both would be replaced by (at least the option of) artificial reproduction: children would be born to both sexes equally.'[31] As mentioned earlier, Firestone's vision effectively exposes an important aspect of feminist debate in this area, namely, that equality does not demand the abolition of natural pregnancy, but rather the *option* of an alternative for those women who would seek it. Smajdor further emphasises that the unequal distribution of risks and burdens involved in the creation of offspring often carries over into inequitable distribution of childrearing responsibilities, thereby feeding into 'assumptions about women's roles as mothers, which restrict and thwart women's ability to function as men's equals in society.'[32] I would therefore argue that disrupting the gestational bonds between mothers and their biological children, which have hitherto been taken for granted, is likely to cause a social re-evaluation of the parenting role beyond early development. By eliminating the necessity for women's temporary incapacitation through pregnancy, ectogenesis could also provide a means of procreating that prevents many women from slipping into the regressive social norms governing childrearing. Although pursuing ectogenesis would not erase the issues of gender inequality, domestic violence or poor maternal care, I agree it is nonetheless a necessary step toward equal opportunity between men and women in reproduction. This is because ectogenesis represents the *only* possible method of producing genetically related children without the need for either parent to be pregnant or to engage a surrogate.

Achieving equality between men and women in reproduction is not only expected to benefit women, as there are many possible ways in which men could also benefit from a more equal role in procreation. As this book adopts a feminist perspective these are not discussed in detail here, but suffice it to say that assumptions about single and gay fathers and stay-at-home dads are likely to be challenged by the advent of ectogenesis technology. Colker claims that men are 'socialized to overvalue their sperm, because their sperm is their only connection to reproduction,' whereas women often have both genetic and gestational ties to their offspring.[33] It would be impolitic to assume that all men are satisfied with such a comparatively diminished role.

Promoting equality between women

Ectogenesis could also reduce the impact of several inequalities between fertile, infertile and homosexual women, with the first currently the only category able to reproduce without first overcoming significant barriers (assuming a willing partner can be acquired!). Owing to pronatalistic social pressures many women in Australia and other industrialised countries feel their feminine identity is linked to their fertility.[34] As such, infertility is likely to be viewed as a serious misfortune, with Anne Donchin claiming women in such circumstances are often socialised to 'look upon their barrenness as a mutilation.'[35] However, from a feminist perspective there is more at stake than mere socialisation, as many women who cannot successfully achieve a pregnancy *genuinely* desire genetic offspring. That these women are often left with no other option than to submit to the highly invasive, and often unsuccessful process of IVF to produce biologically related children, demonstrates unfair disadvantage compared with women able to achieve natural pregnancy. Given that IVF-assisted reproduction is most likely to fail at the implantation stage, ectogenesis could serve to assist such women in pursuing their own biological offspring.[36] According to Rowland, 'the right of infertile women to have children is as imperative a right as that of being childfree.'[37] However, while contraception and family planning have made the latter lifestyle choice a possibility for many women, the former is still subject to brute luck. There is also significant inequality in reproductive success among women on the basis of age, with fertility steadily declining over the lifespan, including through ARTs. While IVF is sometimes criticised because it does not actually 'cure' infertility, it is often justified therapeutically on the grounds that it gives women an opportunity to achieve their end goal of producing biological children.[38] A similar argument would hold for ectogenesis, that although it may not be achieving true equality between the physiologically fertile and infertile, it would at least be addressing the major consequence of this inequality. Neither the successful implementation of IVF programs nor the development of ectogenesis services replaces the need for continued research into genuine cures for infertility.

According to some feminist writers, the mere referral to fertility services can be potentially coercive for infertile women, in that they offer the only possible way of achieving the socially acceptable goal of biological parenthood.[39] Although ectogenesis would not alleviate this particular concern, it could minimise the physical and emotional strain associated with these interventions, for example, by removing the potential for repeated spontaneous abortion resulting from unsuccessful embryo transfer. This is particularly valuable as infertile women often cite multiple miscarriages as one of the worst implications of failed IVF treatment.[40] Research indicates that infertile women often blame themselves for their infertility, looking for past indiscretions that might account for their present misfortune, such as 'risky' sexual behaviour, previous terminations, dietary inadequacies or a variety of other lifestyle factors often not causally related to their infertility.[41] In a pronatalist society where many social structures are built around the concept of the nuclear family, a diagnosis of infertility can lead some women to obsessively pursue alternative methods of procreating in order to fulfil this social expectation. Ectogenesis could assist many infertile women to produce biological offspring, and may be the only viable method for women who have undergone a hysterectomy. A woman without a uterus is deemed to have 'absolute' infertility, because there are currently no ARTs that can help her achieve a pregnancy.[42] Many such women either do not wish to or are prohibited by law from engaging a surrogate. Thus, promoting ectogenesis research could serve to enhance equality between fertile and infertile women in striving to achieve the same goal of biological motherhood, even when the physiological requirements for conception or gestation are not met.

It is not just infertile women who may benefit from the potential for ectogenesis, but also women who are fertile but simply do not wish to expose themselves to the risks and burdens of pregnancy. As demonstrated in earlier sections, regardless of any improvements in maternity care, the health risks of pregnancy and childbirth will always far exceed those of normal everyday life, thereby justifying women's claims to alternative technologies for human gestation.[43] It is often noted in pro-choice literature that while adoption may be the alternative to parenthood, abortion is the alternative to pregnancy. What I believe is exposed in this statement is that there is still a subset of the population left with very limited options on discovering an unwanted pregnancy. For women who do not wish to carry a fetus to term, but who also morally object to abortion, ectogenesis could provide a necessary third option: fetal transport to an artificial uterus for incubation and subsequent adoption. This is particularly important when considering that current contraceptive methods are never completely reliable and that the consequences of failed contraception affect women far more than men. Furthermore, even if the efficacy of contraceptives were greatly improved, there would still be a need to provide options for women who are the victims of rape.[44] Ectogenesis could assist those victims who personally oppose abortion, but nonetheless do not want to continue a pregnancy that was forced on them by an act of violence. While preliminary research into women's attitudes toward ectogenesis indicates that at least initially the uptake level will be minimal, the examples given demonstrate that there is a market for this technology, the needs of which are not being met by current alternatives.[45]

As in the case of male reproduction, through ectogenesis women could supply their own gametes and then allow the growth and development of the fetus to occur outside their body, experiencing no alteration in their usual routine as a result. Smajdor notes that such a choice can be 'defended against charges of being mere whim, preference, or expensive taste' due to the very real health risks associated with pregnancy and childbirth.[46] While much of the opposition toward emerging ectogenesis technologies actually comes from women, equal opportunity dictates that even if the majority of the female population are not interested or find ectogenesis morally objectionable, this does not mean they have the right to prevent other women from accessing something they may consider personally beneficial. This argument carries the most weight when considering women who have no other means of procreating except through the use of new ARTs; however, it also applies to women whose personal preferences may not happen to align with the majority.

Promoting equality irrespective of sexual orientation

Another group poised to benefit from the potential for ectogenesis technology is the gay and lesbian community. Although my major goal in this book is to present a liberal feminist argument for ectogenesis, the rights of single and homosexual men do factor meaningfully in the ART debate. While some single and lesbian women currently benefit from sperm donations, John N. Edwards notes that before ectogenesis becomes a reality, gay men will still be entirely dependent on surrogates if they wish to procreate.[47] Given that in many countries, including Australia, surrogacy is very tightly regulated, and these regulations often expose certain discriminatory attitudes toward same-sex couples, many gay men are automatically excluded from this important aspect of human life. Many bioethicists believe reproduction should be considered a basic human right, with some even advocating for a positive 'right to procreate'.[48] Although such arguments are often used to defend access to ARTs for the physiologically infertile, socially infertile individuals are often ignored. Interestingly, in places where IVF services are limited to married couples, the fact that many of these people are individually fertile is not considered grounds to deny access, as the couple are considered infertile *as a unit*. Presumably the same argument could be advanced to provide access to all socially infertile couples and singletons. As Simonstein and Mashiach-Eizenberg note, one of the major benefits of ectogenesis would be its potential to serve the needs of those who 'lack a womb (for any reason)'.[49] Treating gay couples as a single fertility unit would therefore place their reproductive technology needs on equal footing with heterosexual couples in which the woman has undergone a hysterectomy.

For both heterosexual and homosexual couples, ectogenesis also carries with it the possibility of truly equal biological contributions to the fetus, particularly if ovular merging and parthenogenesis become feasible options for lesbian couples.[50] Dr Orly Lacham-Kaplan's research into spermless reproduction promises to yield interesting results in this area, and is of particular importance when considering that lesbian women often enjoy far fewer legal protections when sperm donors later seek parental recognition for their genetic progeny.[51] The arguments in favour of promoting technological alternatives to physical gestation for lesbian couples are identical to those for heterosexual women, regardless of fertility status. Lacham-Kaplan's research also suggests that two male partners could contribute genetically to a child, with the sperm of one partner being used to fertilise a donor egg that has had all maternal DNA removed and replaced with the nucleus of one of the other male partner's somatic cells. Although single and gay men are still likely to rely on ova banks to provide the necessary materials for procreation, at least in the foreseeable future, the removal of the need for a gestational mother and possible surrogacy arrangements can still be viewed as a significant advancement toward equal opportunity in reproductive choices. In the 1978 article 'Forgotten Fathers' for the gay liberation journal *The Body Politic*, it is lamented that because men cannot gestate their children, all gay fathers have 'a mother to deal with: wife, lover, or ex-wife'.[52] It is also noted that many gay couples achieve a more equitable division of childrearing responsibilities simply because they do not have to first overcome the patriarchal assumptions regarding the sexual division of labour in childcare that impact heterosexual relationships. With the ever-increasing number of gay couples in countries like Australia seeking the assistance of surrogates to achieve their goal of starting a family, and with the recent exclusion of gay couples from engaging international surrogacy agencies in India and Thailand, there is good reason to believe ectogenesis would provide a welcome alternative.[53] Further down the track, stem cell research could even yield an alternative to donor eggs for gay couples, creating embryos entirely from their own genetic material.[54]

When evaluating the claim that men should have equal right to be *pregnant*, rather than just to biologically reproduce, bioethicist Robert Sparrow notes: 'All men currently suffer from a tragic form of infertility – "male barrenness"! It is true that the vast majority of men are not troubled by this sad state of affairs.'[55] However, Sparrow does identify two specific groups likely to suffer substantially by this lack of physical capacity for pregnancy: gay men and male-to-female transsexuals. While research into whether such individuals could potentially achieve safe extrauterine pregnancies has not been encouraging, ectogenesis may provide the next best thing for the wombless, whether male or female.[56]

Promoting equal opportunity in employment

Having discussed where inequality may exist in reproductive endeavours, I now wish to explore the impact this has on opportunity, particularly women's career opportunities. Although there are various equal opportunity issues that could be discussed here, employment is one of the more tangible examples, as instances of pregnancy-based discrimination are common in the literature regarding employment inequality. In 2013, the Fair Work Ombudsman in Australia reported that for the first time since records began, pregnancy discrimination complaints outnumbered disability discrimination complaints.[57] It is well known that women are often under-represented in senior professional roles, including political and corporate leadership positions, and European studies conducted over the past decade have indicated that over half of the women who *do* attain higher-grade managerial positions choose to remain childless.[58] This indicates that time taken out of paid employment for childbearing has a significant impact on women's future career trajectories. According to David Conway, women may be 'unfairly handicapped' in competition for higher-level jobs, due to the 'incapacitating' events of pregnancy and childbirth necessitating breaks in women's continued length of service.[59] He further notes that traditionally the burden of infant care has fallen to biological mothers, further limiting their capacity to engage in paid employment. Conway contrasts these mothers, who are obliged to temporarily withdraw from their occupations, with new fathers, whose dedication to full-time work is significantly less affected by the arrival of children. Hirshman posits that this inequality represents a 'sex-specific brain drain,' which she argues is particularly problematic when highly educated women are being removed from positions of social authority, in order to fulfil non-specialist domestic duties.[60] She claims that professional women who abandon the workplace to become stay-at-home mothers are not 'using their capacities fully,' and that this has a negative impact on both the individual woman and the wider community.[61]

Recognising that women have the right to engage their talents wherever they find the most satisfaction, I propose no conflict with the decision of many professional women to prioritise childrearing over certain occupational pursuits. I am merely suggesting that this sacrifice should not be *expected* of any woman just *because* she is a woman. Instead, a critical assessment of the reasonable distribution of childrearing responsibilities should be undertaken in any reproductive endeavour, in which both partners have equal negotiating power and there is no baseline assumption that the woman should be the one to put her career on hold. The potential for ectogenesis to remove the incapacitating events of pregnancy and childbirth would go a long way to securing this negotiating power. It is important to remember that even in cases where a woman's partner intends to be the primary caregiver of their child, or they mean to access childcare or hire a nanny, the physical burdens of procreation still currently impose unequal restrictions on the woman's employment. Ectogenesis could, therefore, make it more feasible for women in high-power careers to have children by minimising the time away from paid employment required, or in some cases eliminating it altogether.[62]

In addition to interrupting women's continued service, there have also been many cases when women's reproductive capacity has been deliberately used as a method of excluding women from certain professions. Examples include women being fired from positions in laboratories or lead factories after it was discovered certain chemicals could have a teratogenic effect. In 'No Fertile Women Need Apply,' Jeanne Stellman and Mary Sue Henifin recount various inconsistencies in how large companies have historically responded to information regarding hazardous chemicals, often exposing distinct gender-based discrimination. They note, for example, that when it was discovered lead exposure could harm the fetus, women were excluded from employment in lead factories, but not from traditionally 'female' occupations, such as pottery painting, despite the fact that lead was used in these workplaces as well. They discuss how mandatory pregnancy tests and menstrual cycle reporting were used to ensure women were not unknowingly pregnant before coming into contact with potentially harmful chemicals, showing a lack of respect for employee privacy. Furthermore, whenever new research indicated men's reproductive health was at risk from a given chemical, rather than excluding men from employment, industries frequently discontinued production (as in the case of the soil fumigant known as DBCP), or solved the issue by lowering the level of exposure employees were subjected to (as in the case of cadmium and vinyl chloride).[63] While relevant industry representatives have gone on record stating gender-based employment discrimination was done out of concern for 'potential fetuses,' Stellman and Henifin ask: 'what are "potential fetuses" other than an invention that converts women's childbearing capacity into an imaginary permanent state of pregnancy?'[64] This form of discrimination led some women who worked in lead factories to opt for sterilisation rather than lose their only means of support, demonstrating the destructive nature of treating all women solely on the basis of their reproductive capacity.

One problem in establishing that these forms of workplace discrimination constitute sex discrimination relates to the fact that only women can become pregnant. As Colker states, 'The task of determining how men would be treated if they could become pregnant is extremely difficult, since it is a counterfactual enquiry.'[65] She further claims that in the United States the Supreme Court has refused to concede that certain pregnancy-based restrictions on women's employment constitute gender-based discrimination as, 'all pregnant *people* are treated alike; it is irrelevant (to the Supreme Court) that all pregnant people are women.'[66] Thus certain workplaces can exclude pregnant or potentially pregnant *people* on the grounds of avoiding fetal damage, without being held accountable for deliberately excluding *women* from their professions. There are also various occupational positions women simply cannot hold while pregnant, including those that are overly physically demanding or require a particular physique, such as modelling or professional athletics.

It is evident from these examples, and those already discussed in the Introduction, that pregnancy and childbirth have the potential to materially damage women's earning capacity and future job prospects. While in a pronatalist society becoming a mother may be considered a 'promotion to full adult female status,' Oakley notes that it almost invariably entails a demotion in terms of present and future employment.[67] Time spent away from work due to the demands of pregnancy, childbirth and childrearing is likely to affect women's attainment of long-term career goals. Ectogenesis has the potential to remove the first two of these constraints, and spark a re-evaluation of the unequal distribution of domestic labour often implied in the third. However, despite the benefits of ectogenesis in terms of promoting equality and preventing pregnancy-related morbidity and mortality, many ethicists are concerned that the technology would be misused if not regulated properly.[68] The next chapter will explore how equal opportunity must be protected from any potentially negative impact new ARTs, such as ectogenesis, may have on society and women's health.

Notes

1. Smajdor, 'The Moral Imperative for Ectogenesis,' 340.
2. Anna Smajdor, 'In Defense of Ectogenesis,' *Cambridge Quarterly of Healthcare Ethics* 21, no. 1 (2012): 90.
3. David Reuben, 'Organizational Interventions to Improve Health Outcomes of Older Persons,' *Medical Care* 40, no. 5 (2002): 418.
4. Firestone, *The Dialectic of Sex*, 10.
5. Helen H. Lambert, 'Biology and Equality: A Perspective on Sex Differences,' *Signs* 4, no. 1 (1978): 117.
6. Smajdor, 'In Defense of Ectogenesis,' 91.
7. Pence, 'What's So Good about Natural Motherhood?' 80.
8. American College of Obstetricians and Gynecologists (ACOG), 'Pain Relief during Labor,' *Committee Opinion*, no. 295 (2004, reaffirmed 2008): 1.
9. Firestone, *The Dialectic of Sex*, 197.
10. Margaret G. Spinelli, 'Postpartum Psychosis: Detection of Risk and Management,' *American Journal of Psychiatry* 166, no. 4 (2009): 406.
11. Ibid.

12 Smajdor, 'In Defense of Ectogenesis,' 91.
13 ACOG, 'Pain Relief during Labor,' 1.
14 Timothy F. Murphy, 'The Ethics of Impossible and Possible Changes to Human Nature,' *Bioethics* 26, no. 4 (2012): 196.
15 Smajdor, 'The Moral Imperative for Ectogenesis,' 341.
16 Oakley, 'Gender and Generation,' 27.
17 Colker, *Pregnant Men*, 137.
18 Singer and Wells, *Making Babies*, 59.
19 Colker, *Pregnant Men*, 150. In the famous *Davis* vs. *Davis* case in Tennessee in 1990, the male partner's right *not* to be a genetic parent was given priority over his ex-wife's desire to use their IVF embryos to have a child, despite the additional pain Mrs Davis suffered to create these embryos (Samuel A. Gunsburg, 'Frozen Life's Dominion: Extending Reproductive Autonomy Rights to *In Vitro* Fertilization,' *Fordham Law Review* 65 [1996–7]: 2213).
20 Colker, *Pregnant Men*, 150.
21 Heriot, 'Fetal Rights versus the Female Body,' 190.
22 Ibid.
23 Ibid.
24 Ibid.; Colker, *Pregnant Men*, 92. Although data is more difficult to gain, the WHO indicates a similar pattern of violence is experienced by women in the developing world (Kajsa Åsling-Monemi, Rodolfo Peña, Mary Carroll Ellsberg and Lars Åke Persson, 'Violence against Women Increases the Risk of Infant and Child Mortality: A Case-Referent Study in Nicaragua,' *Bulletin of the World Health Organization* 81, no. 1 [2003]: 10–18).
25 Sander-Staudt, 'Of Machine Born?' 116.
26 Ibid.
27 Colker, *Pregnant Men*, 105.
28 Sheila C. Hunt and Ann M. Martin, *Pregnant Women Violent Men: What Midwives Need to Know* (Oxford: Butterworth-Heinemann, 2001), 16.
29 One stark illustration is the 1986 legal case from California *People* vs. *Stewart*, in which Pamela Rae Stewart was charged with criminal child neglect for failing to heed medical advice while pregnant, including refraining from drugs and sex. Despite participating in the sex, her husband was never charged (Joan C. Callahan and James W. Knight, 'Women, Fetuses, Medicine and the Law,' in *Feminist Perspectives in Medical Ethics*, eds Helen Bequaert Holmes and Laura M. Purdy [Bloomington: Indiana University Press, 1992], 235).
30 Peter Singer and Deane Wells, 'Ectogenesis,' in *Ectogenesis: Artificial Womb Technology and the Future of Human Reproduction*, eds Scott Gelfand and John Shook (New York: Rodopi, 2006), 14.
31 Firestone, *The Dialectic of Sex*, 12.
32 Smajdor, 'In Defense of Ectogenesis,' 91.

33 Colker, *Pregnant Men*, 143.
34 Marilyn Anderson, 'Fertility Futures: Implications of National, Pronatalistic Policies for Adolescent Women in Australia,' in *Proceedings of the International Women's Conference, Toowoomba, Queensland, 26–29 September 2009* (Queensland: University of South Queensland, 2007), 40.
35 Anne Donchin, 'The Future of Mothering: Reproductive Technology and Feminist Theory,' *Hypatia* 1, no. 2 (1986): 126.
36 Simonstein and Mashiach-Eizenberg, 'The Artificial Womb,' 88.
37 Rowland, 'Technology and Motherhood,' 516.
38 Jones, *Brave New People*, 105.
39 Meyers, 'The Rush to Motherhood,' 759.
40 Although many feminists rightly find the term 'miscarriage' objectionable because it implies blame for women whose fetuses do not naturally develop to viability, I have chosen to retain its use here, as it is the word most often used by women interviewed on the subject.
41 Carolyn McLeod and Julie Ponesse, 'Infertility and Moral Luck: The Politics of Women Blaming Themselves for Infertility,' *The Journal of Feminist Approaches to Bioethics* 1, no. 1 (2008): 133.
42 Maili Malin Silverio and Elina Hemminki, 'Practice of *In-vitro* Fertilization: A Case Study from Finland,' *Social Science and Medicine* 42, no. 7 (1996): 982.
43 Smajdor, 'The Moral Imperative for Ectogenesis,' 340.
44 It should be noted here that some affected individuals prefer the term 'survivor' while others prefer 'victim.'
45 Cannold, 'Women, Ectogenesis, and Ethical Theory,' 55.
46 Smajdor, 'The Moral Imperative for Ectogenesis,' 340.
47 John N. Edwards, 'New Conceptions: Biosocial Innovations and the Family,' *Journal of Marriage and Family* 53, no. 2 (1991): 355.
48 Paul Lauritzen, *Pursuing Parenthood* (Indiana: Indiana University Press, 1993), 4.
49 Simonstein and Mashiach-Eizenberg, 'The Artificial Womb,' 88.
50 Warren, 'Making Babies,' 289.
51 Deborah Josefson, 'Scientists Fertilise Mouse Eggs without Sperm,' *British Medical Journal* 323, no. 7305 (2001): 127; Colker, *Pregnant Men*, 138.
52 Michael Lynch, 'Forgotten Fathers,' *The Body Politic: Gay Liberation Journal*, April 1978, 10.
53 Julia Medew, 'Surrogacy for Gay Couples in Victoria, Australia,' *The Sydney Morning Herald*, 7 March 2015. Available at: http://www.smh.com.au/national/health/surrogacy-for-gay-couples-in-victoria-australia-20150306-13xcd5.html
54 Catherine D. Payne, 'Stem Cell Research and Cloning for Human Reproduction: An Analysis of the Laws, the Direction in Which They May Be Heading in Light of Recent Developments, and Potential Constitutional Issues,' *Mercer Law Review* 61 (2010): 948; 952.

55 Sparrow, 'Is It "Every Man's Right to Have Babies If He Wants Them"?' 291.
56 Ibid., 276.
57 Lucy Carter, 'Pregnancy Overtakes Disability as Top Source of Workplace Discrimination Complaints,' *ABC News*, 6 November 2013. Available at: http://www.abc.net.au/news/2013-11-06/pregnancy-overtakes-disability-as-the-top-source-of-discriminati/5072904.
58 Catherine Hakim, *Work-Lifestyle Choices in the 21st Century: Preference Theory* (Oxford: Oxford University Press, 2000), 244.
59 David Conway, 'Do Women Benefit from Equal Opportunities Legislation?' in *Equal Opportunities: A Feminist Fallacy*, ed. Caroline Quest (London: Goron Pro-Print, 1992), 53.
60 Hirshman, *Get to Work*, 10.
61 Ibid., 2.
62 When addressing feminist concerns about ectogenesis related to the dominance of males within the reproductive medicine profession, it must be recalled that the under-representation of women within such specialist fields is often a reflection of the competing demands of family and career that women face. That most female doctors choose to become general practitioners, rather than specialists in Australia, is often believed to be due to the fact that the time required to train and qualify in these fields often conflicts with women's family plans (Australian Government Health Workforce Australia, 'Australia's Health Workforce Series: Doctors in Focus 2012' [Adelaide: Health Workforce Australia, 2012], 12).
63 Jeanne Stellman and Mary Sue Henifin, 'No Fertile Women Need Apply: Employment Discrimination and Reproductive Hazards in the Workplace,' in *Biological Woman – The Convenient Myth: A Collection of Feminist Essays and a Comprehensive Bibliography*, eds Ruth Hubbard, Mary Sue Henifin and Barbara Fried (Cambridge: Schenkman Publishing, 1982), 117–20.
64 Ibid., 121.
65 Colker, *Pregnant Men*, 128.
66 Ibid.
67 Oakley, 'Gender and Generation,' 29.
68 Edwards, 'New Conceptions,' 358.

2
Protecting Equal Opportunity from Ectogenesis

Abstract: *Kendal addresses some of the leading feminist concerns regarding assisted reproductive technologies (ARTs), including that they are potentially coercive for infertile women. This chapter also discusses pregnancy discrimination, abortion, fetal rights and disability discrimination, ultimately concluding that none of these concerns represents an insurmountable obstacle for the ethical pursuit of ectogenesis research.*

Keywords: abortion; disability discrimination; fetal rights; pregnancy discrimination; severance theory

Kendal, Evie. *Equal Opportunity and the Case for State Sponsored Ectogenesis*. Basingstoke: Palgrave Macmillan, 2015. DOI: 10.1057/9781137549877.0006.

It is often lamented in bioethics literature that new technologies are not sufficiently scrutinised before release, with some arguing that now is the time to discuss how ectogenesis might affect women and society.[1] The rapid development of IVF and embryo transfer techniques suggest full ectogenesis may not be far from being realised, and thus the positive and negative outcomes of this technology should be considered with some urgency in order to protect the interests of all parties involved. While promoting equal opportunity *through* ectogenesis requires that this new technology be supported as one of a number of possible methods for dealing with existing inequalities in human reproduction, protecting equal opportunity *from* ectogenesis relies on establishing a baseline of rights and opportunities below which no one will fall as a direct result of embracing this technology. For this argument I am using existing medical and legal standards as the baseline, not as a defence of this status quo, but rather as representing a minimal threshold of rights for women and fetuses. When discussing whether providing additional choices in reproductive medicine is actually of benefit to women, Gerald Dworkin uses the theory of Pareto optimality to argue that as a 'normative criterion' more choice is better only if 'someone is made better off and no one is made worse off.'[2] In this economic theory a Pareto improvement occurs only when an advantage is gained for one party without concomitant disadvantage to others, with the optimal state being the point at which no further changes can be made to the system without violating this principle. In Chapter 1, various arguments were presented that could satisfy the first of these requirements. This present chapter will now focus on satisfying the second, by discussing how equal opportunity can be protected in the face of increasing technological interference in the human reproductive process. As embracing ectogenesis would also involve the allocation of significant resources, establishing the need for any redistribution of assets to form a Pareto improvement helps defend supporters of this technology from accusations of catering to the needs of some at the expense of others.

In this chapter I will be exploring some potentially deleterious effects of ectogenesis and suggesting means by which possible threats to equal opportunity could be reduced or eliminated such that no individual is disadvantaged by the release of this technology. Such measures include protecting current pregnancy anti-discrimination laws and maternity leave arrangements, respecting the right of women to refuse this new technology, supporting the right of women to continue pursuing family through alternate means, including adoption, and discussing how ectogenesis may influence the abortion debate. The final point that will be briefly covered is whether there is a case for prohibiting ectogenesis on the basis of fetal harm or lack of consent. It will be my contention that although there are significant challenges to the ethical provision of ectogenesis services, these are not insurmountable and thus do not justify prohibiting a technology that could make a considerable contribution to improving the situation of many women.

Protecting against pregnancy discrimination

The first element I wish to discuss is the need to protect women's freedom to pursue physical pregnancy, even after alternatives such as ectogenesis become available. This is particularly important when considering the hard-won battles for maternity leave arrangements and pregnancy anti-discrimination laws that are currently in place to protect equal opportunity for women. From a feminist perspective, it is necessary to establish that the right to exercise choice among some women does not serve to disadvantage others. Thus, catering to the preferences of some people to reproduce using ectogenesis should not be used to erode the rights of women who still wish to experience physical pregnancy, including their right to work-based maternity leave arrangements.

The history of pregnancy anti-discrimination law is relevant to the ARTs debate, particularly when considering a new technology that could remove the need for some women to be physically pregnant. Nina G. Golden reminds us that 30 years ago women could be summarily dismissed from their jobs merely for announcing a pregnancy with no consequences for their employers. Furthermore, she notes that 20 years ago there was still no job protection for women who wanted to take time off to care for newborn children.[3] Although the situation has certainly improved since then, there is still cause for considerable concern regarding pregnancy discrimination. In response to empirical evidence that women were being disadvantaged in the workplace due to reproductive difference, many industrialised countries created legislation aimed at eliminating pregnancy-based discrimination and increasing female participation in the workforce.[4] This has been identified as a particular priority in Australia due to the economic strain of an ageing population, exacerbated by declining fertility rates.[5] Pregnancy discrimination is covered in the United Kingdom under the *Sex Discrimination Act* (1975) and *Pregnant Worker Directive* (1992), in the United States by the *Sex Discrimination Act* (1978) and *The Family and Medical Leave Act* (1993) and in Australia by the *Sex Discrimination Act* (1984). Both the UK and Australian Acts establish a requirement for workplaces to make reasonable accommodations for pregnant workers and contain a 'right-to-request' policy regarding flexible work arrangements during pregnancy.[6] Meanwhile the US federal policies forbid discrimination against pregnant women, but require no special treatment to meet their unique needs at this time.[7] Some US state policies, such as the *California Family Rights Act* (1991), go further to provide specific pregnancy leave options for workers affected by pregnancy-related illness, in addition to state-funded paid maternity leave.[8] However, even this Act, known as the most 'generous' in America, does not guarantee job security for women with salaries in the top 10 per cent of any workplace, introducing unfair risk to the few women who do manage to achieve seniority in employment.[9] While such policies ensure that many women are guaranteed their former positions back when they return from maternity leave, these exceptions are particularly troubling when considering the under-representation of women in elite jobs.

Most US states do not provide paid maternity leave and until very recently the same was true of Australia.[10] Also, despite the aforementioned legislation, there are still many cases of pregnancy discrimination documented in the United States, United Kingdom and Australia, with a high probability of under-reporting. Nancy Casas claims that in Australia an 'alarming number' of employers continue to breach the *Sex Discrimination Act* (1984), either intentionally or due to a lack of knowledge regarding their obligations to pregnant employees.[11] This discrimination can be both direct, involving demotion, dismissal and exclusion from training opportunities, and indirect, involving a lack of accommodation for pregnant women or considering maternity leave as an interruption to continued service. In 'Expecting the Worst,' Paula McDonald et al. discuss a litany of discriminatory offences perpetrated against pregnant employees in Australia, including subjection to derogatory comments (particularly about their physical appearance), unfavourable performance appraisals (especially from male managers) and changed work hours or conditions. Such offences are often committed with the express intention of coercing pregnant employees into resigning their positions. Reports of such discriminatory workplace behaviour are particularly common among groups of women employed in less privileged occupations, and are most likely made worse by the high proportion of women in casual employment in Australia.[12]

As a result of concerns regarding pregnancy-based discrimination, the Equal Opportunity and Human Rights Commission (HREOC) produced a report in 1999 that concluded the federal government needed to do more to protect the rights of pregnant workers.[13] Among the more common violations found were employers refusing to make reasonable adjustments for pregnant women, such as providing seating, adequate toilet breaks and access to drinking water. Although US federal law does not require such accommodations, Australia's anti-discrimination policy *does* stipulate a requirement to take reasonable measures to ease workplace strain for pregnant employees. According to Casas, the HREOC's report 'found that the stereotyping of a woman's capacity to work while pregnant leads to discriminatory dismissals and employer-imposed limitations on women in the workplace.'[14] McDonald et al. further claim that cultural stereotypes regarding how women should behave while pregnant, and negative reactions to pregnant bodies functioning in public roles, also bias employers against pregnant employees, leading to discriminatory behaviour.[15] At the time of writing the president of the Australian Human Rights Commission (formerly HREOC), Gillian Triggs, has noted a 'growing number of complaints about pregnancy discrimination,' describing these numbers as far worse than anticipated.[16] She has announced that she wishes the commission to focus on such 'mainstream' issues in employment discrimination into the future.

While many of the stereotypes that contribute to pregnancy discrimination could be challenged by the advent of ectogenesis technology, the examples provided demonstrate that better protection is needed for women choosing physical pregnancy *before* this alternative becomes available. Even for women and couples using ectogenesis, provisions for parental leave will still be necessary to allow for the care of newborns and should remain firmly protected. However, as many women will still prefer physical pregnancy, the demands of equal opportunity require that we rigorously enforce existing anti-discrimination laws and increase employer and employee education regarding the rights of pregnant women in the workplace. This is necessary to prevent undue pressure on working women to use ectogenesis rather than inconvenience their employers. Just as it is vitally important that pregnant women be granted equal opportunity in the workplace compared with men, so must there be legislation in place to ensure they are not disadvantaged compared with women who choose to reproduce via ectogenesis. Furthermore, while achieving a Pareto improvement only requires that the introduction of ectogenesis not erode the current rights of physically pregnant women in its goal of improving options for other women, I argue that far more needs to be done to actually achieve equality for pregnant workers. The continued occurrence of pregnancy discrimination in industrialised countries like Australia indicates that current anti-discrimination laws are not sufficient to protect women from occupational disadvantage associated with their childbearing capacity.

There are also various advocates who argue that infertility discrimination is becoming an increasing issue in the workplace, but one that is not addressed in most anti-discrimination legislation. Jeanne Hayes notes that the demands of IVF can be very disruptive to women's work schedules, but that many are denied the same leave fertile women can access for pregnancy-related conditions. She relates cases where women have been told they have to choose between work and pursuing the goal of pregnancy, particularly as repeated IVF cycles are often necessary to achieve this, thereby extending the period of time workplaces are inconvenienced beyond the predictable nine-month gestational period.[17] Better facilitating the needs of infertile women undergoing IVF would also ensure those women choosing IVF initiation for ectogenesis would be free to access the necessary leave to attend clinical appointments. Again, such rights would need to be negotiated and legally enshrined before this technology was released.

Apart from the physical changes to reproduction that ectogenesis could entail, there are also social ramifications to releasing this technology. Some feminists argue that until these become 'clearly manifest,' the potential risk to certain disadvantaged groups of women justify universal prohibition.[18] However, according to the theory of Pareto optimality, as long as the current situation of these groups is not worsened, the development of ectogenesis can still be supported, as it would benefit others in the system. This does not negate the fact that other support systems are clearly needed to improve the overall health and well-being of disadvantaged women, irrespective of any change in available reproductive options.

As already discussed, ectogenesis has the potential to assist infertile women, while also providing an alternative to both unwanted pregnancy and abortion for women who wish to avoid these. As such, Murphy claims that as long as physical pregnancy remained an option for women and ectogenesis was used only voluntarily, this new technology could avoid condemnation on the grounds of being potentially exploitative.[19] Ectogenesis could also cater to the preferences of women who simply do not wish to experience pregnancy, or who consider it an undesirable state for occupational or health-related reasons. Thus, assuming protections were already in place such that the rights of pregnant women were not jeopardised, ectogenesis could potentially increase choice for some women, without diminishing options for others.

Protecting the right to refuse

In this section I wish to explore in greater depth the need to protect women's right to refuse new technological interventions in the reproductive process, before such interventions become widespread. One of the major criticisms of ARTs is that their mere existence can be considered coercive, particularly for women living in a pronatalist society, such as Australia, in which the goal of motherhood is afforded great cultural significance. The relationship between pronatalist dogma and genetic essentialism is well documented, with the combination exerting pressure on women not only to reproduce, but more specifically, to produce *biologically related* children.[20] Thus, the availability of ectogenesis could be seen as yet another way to promote biological parenthood in preference to other means of starting a family, such as adoption, and to further exacerbate issues surrounding the social disapproval of voluntary childlessness.

While some women currently cite employment goals as the reason they do not wish to become pregnant, ectogenesis could reduce the impact of this 'defence' for their decision to remain childless. This is particularly relevant for nations like Australia that are suffering the economic strain of an ageing population. Pronatalist agendas in such circumstances are often aimed at supporting population growth by manufacturing a perceived social duty for women and couples to procreate, and casting those women who choose career above motherhood as 'selfish and uncaring.'[21] I maintain that the positive goals of ectogenesis, including assisting the infertile, should not be perverted in order to fuel this argument and, furthermore, that the right to remain childless, regardless of fertility status, should be socially embraced before this technology becomes available. In the interests of protecting autonomous decision making for all women in reproductive matters, the right to define family however one may choose is necessary to support the different needs of single and partnered women, heterosexual and lesbian women, biological and adoptive mothers and the voluntarily and involuntarily childless.

According to Cannold, the right to refuse ectogenesis for both fertile and infertile women is particularly important when considering that although ethicists are generally positive towards the technology, 'the results of qualitative research suggest that women's response ... would be overwhelmingly negative.'[22] Thus, promoting ectogenesis alone as the solution to involuntary childlessness fails to address the needs of infertile women who do not wish to rely so heavily on artificial means to become mothers. As such, I argue the provision of ectogenesis must only be *in addition* to whatever options are already on offer for women, including those aimed at establishing a physical pregnancy. Furthermore, there must be no undue pressure on women to try ectogenesis, who would otherwise choose to remain childless or pursue adoption following a diagnosis of infertility. Murphy claims that adoption should in fact be socially prioritised over ectogenesis, for serving the needs of children who already exist, as well as the biological parents who gave them up.[23] Challenging genetic essentialism, this argument defends the choice to refuse technological interference in fertility, while supporting the validity of alternative methods of achieving the goal of family. This is particularly relevant when considering that the advent of IVF is often blamed for the cultural shift in which adoption became 'second best' to technological solutions to involuntary childlessness, pressuring infertile couples to exhaust these options first before abandoning the goal of producing genetically related offspring.[24]

In *Pursuing Parenthood*, Paul Lauritzen claims ARTs already impose 'unwanted choices' on infertile women, highlighting the need for autonomy and informed consent in reproductive decisions.[25] Rather than empowering women with more options, ART service providers have the potential to exploit infertile women, encouraging them to submit to expensive fertility treatments that may be harmful to their health and well-being.[26] Lauritzen argues such interventions often objectify women's bodies and ascribe 'individuality and intentionality' to certain body parts involved in reproduction, such that infertility becomes something that can be blamed on an 'incompetent' cervix or its 'hostile' mucus.[27] Likewise, Martha E. Gimenez argues that the more extreme the artificial interventions imposed on women are, the more fractured their experience of procreation will become, something she calls the 'technological fragmentation of the reproductive process.'[28] According to this logic, since ectogenesis represents an interruption to, or technological replacement of, natural gestation, this could lead to further splintering of the image of the body into its reproductive and non-reproductive components. While the language of reproductive medicine certainly needs to change to reflect greater respect for women and their bodies as a whole, I argue that ectogenesis does not represent a *unique* threat to women's bodily integrity and autonomous decision-making capacity. Consistency would seem to demand that if women can consent to IVF and other ARTs, despite the potential influence of pronatalistic social coercion, there is no reason to prohibit ectogenesis on these grounds. Furthermore, ectogenesis may actually present an opportunity to re-evaluate certain assumptions regarding women's bodies and their roles in procreation, which have historically been taken for granted. Thus, concerns regarding the potentially coercive nature of ARTs need not demand the prohibition of ectogenesis, so long as informed, voluntary and authentic consent is obtained, the right to refuse is protected and other means of starting a family are socially respected.

Lauritzen claims that in cultures where motherhood is valued above any other role a woman may perform, offering an infertile woman the option of remaining involuntarily childless or accessing ARTs is not offering 'genuine choice.'[29] However, as Singer and Wells maintain, any attempt to prohibit ARTs, including ectogenesis, by implying women are not 'responsible agents capable of choosing for themselves' would demonstrate 'extraordinary paternalism – never mind about male chauvinism.'[30] Such infantilisation is already rampant in maternity care, including through the assumption that labouring women are incompetent to make treatment decisions for themselves regarding pain relief.[31] Although I agree with Lauritzen that pronatalistic social coercion certainly *can* diminish autonomy, I argue that this is grounds only to re-evaluate pronatalist dogma itself, *not* to prohibit new ARTs. Indeed, the suggestion that women are incapable of withstanding social pressure is different to claiming that they simply should not have to. This includes current pressures to become pregnant through natural or assisted means, and possible future pressures to access ectogenesis. Respect for autonomy requires that all concepts of 'the family' be treated equally, including those that embrace biological parenthood, adoption, childlessness and technologically facilitated parenthood. As autonomy cannot be maximised through imposing an absence of choice, the argument that the existence of new technologies is potentially coercive and therefore should be prohibited is not sound. It must also be noted that at present all women 'choosing' to be physically pregnant to produce biologically related offspring are doing so in the absence of a viable alternative.

Another consideration when protecting the right to refuse is the increasing medicalisation of pregnancy and childbirth, which has led to the normalisation of technological intervention in the childbearing process. The history of reproductive medicine demonstrates this normalising effect, sometimes leading to a failure to recognise that medical and technological interventions in procreation are still subjects of choice. Ultrasonography is one such technology that has been subsumed into routine pregnancy management, with one study conducted in the mid-1990s finding the majority of women recruited from maternity clinics in the United Kingdom were not aware that scanning was optional, as it was merely accepted as standard and informed consent was not sought.[32] The same was found in another study regarding interventions used during childbirth in Western hospitals, such as episiotomies and oxytocin use, revealing that women were not being presented with all the options necessary to make an informed choice.[33] Gimenez claims that rather than empowering women with more choice, the over-medicalisation of pregnancy and childbirth relegates women to the position of passive victims of 'pharmacrats' and 'technodocs.'[34] Elizabeth Heitman further notes that the medicalisation of pregnancy and childbirth is particularly seductive for infertile couples, as it provides a 'morally neutral explanation' for their stigmatised state.[35] I argue this perceptual shift from regarding pregnancy and childbirth simply as a naturally occurring life event to treating it as a condition requiring hospitalisation and extensive medical intervention may lead some women to consider procreation a medical, rather than natural, phenomenon. Although this does not necessarily influence autonomous decision making, it does have an impact on the perception of infertility, thereby paving the way for greater uptake of ARTs, potentially including ectogenesis. As women become more accustomed to technological interference in their reproductive lives, ectogenesis for the infertile could one day become just as routine and expected as an ultrasound for pregnant women today. Intuitively, such a future may seem implausible, however, if it were to be realised the 'right to refuse' may effectively disappear, as many women may no longer recognise that there is even a choice at stake. This again highlights the need to obtain informed consent for *all* medical interventions, including those that are already routinely practiced.

It must also be acknowledged that the more technologically advanced reproductive technologies become, the more specialised the knowledge required to understand them. This potential lack of understanding challenges informed consent, particularly if the medicalisation of pregnancy and childbirth attempts to remove the locus of knowledge regarding their own biology from communities of women and resituate it in the predominantly male obstetric profession.[36] Gillian Lockwood notes that the more advanced reproductive technology becomes, the more likely doctors will be to refer their infertile patients.[37] This valorisation of scientific discovery as the solution to human limitations illustrates what Fischer et al. claim is the dominance of 'scientific rationalism' over Western medicine.[38] By reducing infertility to a purely medical problem, the medicalisation of pregnancy and childbirth can lead to a failure to engage with the unique and multifaceted experience of infertility for each individual woman affected, which can be influenced by distinct social and cultural factors. Although the medicalisation of pregnancy may lead more women, both fertile and infertile, to consider ectogenesis who might otherwise choose to pursue physical pregnancy or remain childless, I argue this is not a compelling reason to reject this new technology when other technological interventions are already commonplace in maternity care. So long as the right to refuse is protected regarding these, and ideally all other, technological interventions, the theory of Pareto optimality can be used to defend the allocation of resources to achieve the addition of ectogenesis into the current system.

Importantly, the freedom to pursue a physical pregnancy must stand firm even in cases where it may be medically advisable for a woman to avoid the stress of pregnancy, or assuming a future in which ectogenesis has surpassed physical pregnancy in terms of both maternal *and* fetal safety. As current medical ethics defends the right of patients to refuse even life-saving treatments, the right of women to make 'risky' reproductive decisions is similarly protected.[39] Also, while it is often considered socially unacceptable to engage in activities that are known to harm the fetus, such as smoking and consuming alcohol while pregnant, most feminists oppose regulations that aim to further reduce women's bodily autonomy while in this state. As the fetus is generally not granted full personhood, many feminists argue that whenever a conflict arises between fetal and maternal well-being, the interests of the fetus must remain subservient to those of the woman carrying it.[40] As such, were ectogenesis to become the medically preferred method of procreation, this should not be used to coerce women into abandoning their desire for more traditional reproductive options. The principles of reproductive liberty demand that such choices remain open for women and couples, regardless of whether society at large considers them desirable. The reasons for this are clear when considering the possible methods by which physical pregnancy could be denied to physiologically fertile women, which include forced sterilisation, an abuse that has been recognised by the International Criminal Court as a crime against humanity.[41] Like smoking and drinking while pregnant, it is conceivable that choosing a physical rather than artificial gestation may in the future become socially frowned upon; however, respect for autonomy dictates that none of these activities be forcibly restricted.

Protecting legalised abortion

While it is not my intention in this book to argue for or against abortion rights *per se*, the relationship between ectogenesis and legalised abortion is such that the former cannot be discussed without making some mention of the latter. For the purposes of this section, it will suffice to prove that the successful implementation of ectogenesis technologies need *not* necessarily demand the re-criminalisation of abortion. In this section I will highlight some of the ethical arguments against abortion that might arise once ectogenesis exists, making particular reference to severance theory. While some of these arguments are quite compelling, I will ultimately conclude that legalised abortion is necessary to protect women's health and safety. In terms of protecting opportunity, retaining legalised abortion is in keeping with the idea that current medical and legal standards should represent a baseline for rights in the event that ectogenesis services become available in the future. This does not preclude changes to the current system being initiated on other grounds, such as the growing public sentiment regarding the moral status of the fetus, but suggests that the advent of ectogenesis should not in itself be used to advance anti-choice agendas.

According to severance theory, a woman's right to an abortion is grounded in a respect for bodily integrity, autonomy and privacy, such that she can terminate the unwanted dependence of the fetus on her body, even though this action leads to the death of the fetus as a necessary consequence.[42] For severance theorists, the loss of fetal life is unfortunate but must be tolerated due to the superior moral claims of the woman involved. As Cannold notes, ectogenesis would challenge the assumption that the termination of pregnancy is necessarily 'synonymous with fetal death,' providing a method by which women could end unwanted pregnancies *without* the death of the fetus.[43] According to Cannold, since severance theorists consider fetal death an undesirable, yet heretofore unavoidable, consequence of a woman's right to terminate a pregnancy, this theory 'logically compels its adherents to embrace ectogenesis,' as an alternative that provides for the needs of the woman without sacrificing the interests of the fetus.[44] For this reason, many believe any moral justifications allowing the fetus to be killed on the grounds of respecting a woman's right to control her own body would be invalidated were ectogenesis to become a viable alternative.

In 'Abortion and the Death of the Fetus,' Steven L. Ross outlines the common argument that were the continued existence of the fetus of no danger or inconvenience to its biological mother, there would be no justifiable reason for her to wish the fetus dead. He notes that by this logic, ectogenesis provides for liberal ideals of bodily autonomy for the woman, while also respecting conservative views regarding the fetus as an innocent life that deserves protection. After all, as he states, 'What good reason can we have for wanting *any* complex organism dead when its being alive in no obvious way inconveniences us?'[45] As a woman's right to bodily integrity has traditionally been used as grounds to allow abortion up until the fetus reaches independent viability, ectogenesis could undermine abortion rights as the fetus would *always* be viable with technological intervention.[46] It is often asserted in pro-choice literature that a mother should be 'entitled to refuse to act as a life-support system' for the fetus, but many believe once this demand is no longer in effect, the state may have an interest in preserving the life of the fetus as a separate entity.[47] The viability distinction is not only a problem in the ectogenesis debate however, as this boundary already varies quite substantially depending on geographical location, access to medical technology and many other factors.[48] In fact, in order to avoid any blurring between fetuses that are aborted and those that are induced prematurely but born alive, some regulatory bodies recommend abortive methods that *guarantee* feticide while within the mother's womb, demonstrating that the viability distinction may need to be reconsidered independent of any change to available technologies.[49]

The question now becomes whether abortion serves any other function except terminating the physical dependence of the fetus on the mother's body? What I hope to demonstrate in this section is that while ectogenesis may solve some of the issues associated with an unwanted pregnancy, it does not serve as the *automatic* replacement for abortion. This is because women may seek an abortion for reasons that could not be satisfied simply by providing ectogenesis to take over the task of growing the fetus, and adoption to substitute for parenting responsibilities. Both Ross and Cannold observe that women often express dissatisfaction with ectogenesis as a potential solution to an unwanted pregnancy, indicating that it is not just the physical burden of pregnancy that women are seeking to avoid in requesting an abortion. Ross claims that in many cases it is not that the woman simply desires not to be pregnant, it is that 'they do not want there to *be* a child...They cannot be satisfied *unless* the fetus is killed; nothing else will do.'[50] According to Cannold this is because women may feel a sense of obligation to their genetic offspring, regardless of whether they gestate or raise them, and thus may view abortion 'as a way to *prevent the creation* of something for which, once it comes into being, they will feel inescapably responsible.'[51] I am not suggesting that a woman *should* necessarily have the right to demand the death of the fetus when exercising her right to terminate an unwanted pregnancy, merely that there may be considerations involved in the request for an abortion that ectogenesis would not satisfy. When consistently applying the theory of Pareto optimality, this requires that the addition of ectogenesis into the current system not be accompanied by the removal of current abortion rights, as the existence of the former would not remove all desire for the latter. While it is likely some women would choose this alternative in preference to abortion, others would still wish to avoid becoming genetic mothers altogether, thereby also avoiding the complications associated with adoption.[52] In such cases it would be vital that legal abortions remain an option, to avoid the negative health consequences of unsafe 'backyard' operations.

While it is true that men currently cannot avoid becoming genetic fathers if the biological mother of their offspring chooses to continue a pregnancy against their desires, the fact that the fetus is not just genetically but also materially derived from the mother's body make these two things incommensurable. Respect for bodily integrity demands that a woman not be forced to continue or terminate a pregnancy on the basis of the male partner's genetic relationship to the fetus. Ectogenesis may provide an alternative solution in some cases of parental conflict though, by providing the only possible method by which the biological father could 'keep' the fetus without imposing on the mother.[53] Like all medical procedures, the removal of the fetus from the mother's womb in order to move it to an ectogenesis chamber would have to be agreed to voluntarily. Since women could not ethically be compelled to use ectogenesis, anymore than they should be coerced into continuing an unwanted pregnancy, abortion would remain a necessary alternative to ensure respect for bodily integrity. This argument supports the dominant view that while the fetus is occupying the space within the mother's body, she has the right to determine what happens to it, at least until the point of independent viability.

It is sometimes argued that ectogenesis would provide the only uncontentious method of being both pro-life and pro-choice, as it could allow for the termination of unwanted pregnancies without necessitating the death of the fetus.[54] Thus, ectogenesis represents a necessary technology to cater for the needs of pregnant women who neither wish to carry to term, nor have an abortion. Edwards claims that if full ectogenesis were realised, 'abortions could become obsolete' as they would simply become 'early births' instead.[55] However, as Ross argues, women may request an abortion *specifically* to kill the fetus, in part because this life is not just *any* human life, but rather is their biological child, and thus 'represents one of the potentially most central relationships possible.'[56] According to Ross, it is logically sound for a woman to 'simply want there to be no child at all,' suggesting that while ectogenesis would serve as a welcome solution to the problem of an unwanted pregnancy for some women, it would not function as a substitute for all abortions.[57] Without commenting on the ethics of abortion *per se*, this argument demonstrates that allowing ectogenesis to lead to the re-criminalisation of abortion would leave some women worse off, as it would remove one option without providing a satisfactory substitute that would serve the needs of all women involved. There are grounds for a woman not only to desire a return to her pre-pregnancy physical state, but also to wish for the prevention of the birth of her biological child, especially if she is concerned about it being raised by someone else. Ectogenesis only provides a solution to the first issue, while abortion remains the only sure method of conquering the other objections. Such arguments are convincing regarding the need to protect legalised abortion; however, I argue they do not constitute sufficient grounds to prevent the development of ectogenesis, so long as access to the former was not made contingent on the absence of the latter.

Protecting the rights of the fetus

The other interested party for which ectogenesis must not represent diminished opportunity is the fetus itself. The problem with applying the theory of Pareto optimality to this situation is that it is almost impossible to find a reliable baseline for comparison. The moral and legal status of the fetus is a highly contentious issue, with laws regarding if and when the state has an interest in protecting fetal life often being inconsistent. Two recent examples that have attracted considerable media attention around the world illustrate some of the conflicting ideals of fetal rights existing today. The first is from the United States, with 33-year-old Purvi Patel from Indiana convicted in February 2015 of feticide and child neglect following the death of her 24-week fetus in July 2013. Journalist Nicky Woolf claims this verdict is 'mutually contradictory' as it requires both that the fetus was killed *in utero* and born alive and subsequently neglected.[58] Be that as it may, had Ms Patel been living in various other states of the United States there would have been no law to prosecute her under. Meanwhile, in Sydney, Australia, a pregnant Jehovah's Witness and her seven-month fetus have recently died due to the mother's refusal to undergo a blood transfusion on religious grounds. A caesarean might have produced an independently viable child, but could have hastened the death of the mother and thus was not permitted.[59] While at the time of writing the decision not to save the fetus is being widely criticised, Sascha Callaghan of the University of Sydney notes that priorising the life of the fetus at the expense of its mother's bodily autonomy would be 'diluting' pregnant women's 'full rights as citizens'.[60] For me this highlights the legitimacy of holding the interests of the fetus subservient to that of its mother, as while there is disagreement regarding whether the former should be considered a 'person' in any meaningful sense, there is no doubt that the latter possesses full personhood rights. Nevertheless, when faced with the same dilemma in the United States, a blood transfusion was forced on a non-consenting pregnant Jehovah's Witness in order to save her fetus.[61] Helena Anolak claims that within Australia, the United States and the United Kingdom, court-ordered treatments for non-consenting pregnant women, including forced caesareans, are increasing alongside public debate regarding fetal 'personhood,' with legal systems across the globe being unable to resolve the conflicting views regarding the proper status of the human fetus.[62] It is not just across different states that the treatment of the fetus is inconsistent either, but also within the same state. For example, in Victoria, Australia, babies are being transferred to neonatal intensive care units (NICU) for aggressive treatment aimed at saving their lives when born at gestational ages for which abortion is still legal on demand.

Against this backdrop it is difficult to have a conversation about protecting the rights of the fetus from any threat ectogenesis may pose. In the case of an *unwanted* fetus, the chance of survival as opposed to the certainty of death through abortion would seem to suggest that the needs of such a fetus could be served well by this technology, although there are still ethicists who are concerned about the potential impacts ectogenesis could have on future persons. According to Murphy, the goal of ectogenesis 'must surely be to produce an infant indistinguishable in health and vigor from an infant born of a human pregnancy.'[63] There is insufficient evidence, however, to indicate how artificial incubation, either from conception or shortly after, would affect fetal health and well-being. With regard to the risks involved with fetal transfer from the uterus, Murphy notes that the potential for severe fetal damage might make ectogenesis 'ethically prohibitive.'[64] This is due to a belief that imposing a high risk of disability and injury on a non-consenting future person is unethical. It is important to remember here though that many of the potential risks associated with artificial gestation already exist in natural pregnancy, and that no 'future person' has ever consented in advance to being conceived or born through any method, biological, technological, or otherwise. The intuition that it would be grossly unethical to expose an *ex utero* fetus to teratagenic chemicals, radiation or infections, should at least give us pause regarding whether it might also be unethical to force the same risks on the *in utero* fetus, as the resultant child will bear the consequences of any intentional or accidental exposure to such prenatal hazards. As these threats are already tolerated in natural pregnancies on the grounds that the fetus stands to benefit from the chance of existence, I argue others should similarly be tolerated in the case of ectogenesis. Particularly for unwanted fetuses faced with the alternative of non-existence, I argue that if there are any additional risks associated with artificial gestation these may well be justifiable.

From a justice perspective, it is usually argued that for medical research to be considered ethically acceptable, the risks associated with the research must not be borne by one group of people when the benefits are conferred elsewhere. Thus, the participants for clinical trials must be reasonably expected to benefit from the results of the trial, in order to justify imposing the risks of an unproven treatment on them. Amel Alghrani and Margaret Brazier note that involving fetuses in ectogenesis research may not be in their own best interests, particularly in the earlier stages when success is expected to be minimal. They claim this is especially relevant when considering the risk of severe injury that might affect early test subjects, as there may be a point at which certain death through abortion or non-treatment may be preferable to life with profound disability. However, these authors ultimately conclude that benefit to *future* babies is sufficient ethical grounds to allow ectogenesis research on current fetuses, as '[w]ithout studies like the surfactant trials that offered no real hope of immediate benefit to the neonatal subjects at the time, babies born at what is now seen as a pretty safe time, 28–30 weeks, might not survive today.'[65] As surfactant was the first drug developed specifically for use in premature newborns, adult trials were not a possibility.[66] When considering fetal subjects otherwise destined for abortion, Alghrani and Brazier claim it is 'perverse' to allow a woman to abort with the intention of killing the fetus, but not with the intention of providing it even a slim chance of survival through ectogenesis trials.[67] I agree that maintaining such a position is fundamentally illogical, and that the codes of practice governing fetal research should allow some flexibility in cases where experimental treatments represent the only hope of survival for the fetus.

A more problematic case arises with the intentional creation of a fetus entirely *ex utero*, particularly before the technology is proved safe. As Singer and Wells state, if it is unethical to attempt ectogenesis until we have 'a reasonable assurance that it is safe and we can have no reasonable assurance that it is safe until it has been carried out, we seem to be in a classic "catch 22" situation. Work on ectogenesis will remain forever unjustifiable.'[68] However, Singer also notes that these same concerns were raised regarding IVF, and proved to be unfounded.[69] While some bioethicists draw a distinction between natural reproductive hazards and the supposed 'man-made' risks imposed on the fetus through ARTs, it is difficult to see how biological pregnancy can be considered any less 'man-made' at its core.[70] Regardless, it seems likely that advancements in the field of ectogenesis research may rely on 'serendipitous' births of infants of increasing prematurity, or be initially restricted to use for fetuses which survive abortion. In fact, with regard to the growing public concern about protecting fetal life, I argue that there may come a time when failure to provide access to advanced humidicrib technologies, including ectogenesis, on the basis of extreme prematurity may be considered a form of gestational ageism. Already paediatricians are calling for greater discussion regarding appropriate interventions for babies born at 22 and 23 weeks' gestation, as their survival rates continue to increase when born in hospitals that do not automatically exclude them from receiving treatment.[71]

With the potential for ectogenesis to push the boundary of viability further than ever before, the laws governing fetal rights and the ethics of fetal research will need to be re-examined in order to protect the interests of the fetus. Many laws dealing with reproductive matters are woefully out of date in Australia and other industrialised nations. Despite some revisions, the foundation for many of the laws discussed earlier in this chapter are 20–30 years old, with such advisory documents as the United Kingdom's *Warnock Report* (1978) and the Polkinghorne Committee Code of Practice (1989) also being unequal to the task of dealing with the ramifications of the rapidly advancing field of ARTs. In Australia, relevant documents include the *Research Involving Human Embryos* (2002) and various ethical guidelines released by the National Health and Medical Research Council. Alghrani and Brazier note the age of such documents mean they often '[do] not canvass even the possibility of ectogenesis.'[72] I argue the need to protect the fetus from exploitation in ectogenesis research requires that the relevant laws be updated, such that there is no ambiguity regarding the legal and moral status of the ectogenetic fetus and the limits of its role in experimental research. In the case of prematurely born babies, however, there may be grounds to argue that they should all be granted equal opportunity to reap the potentially lifesaving benefits of artificial incubation, regardless of gestational age or the probability of success.

In addition to challenging the viability distinction for legal abortion, ectogenesis has the potential to drastically alter perceptions of the fetus as a legal entity. Edwards claims the greatest risk associated with new ARTs is that they have 'emerged in what is essentially a legal vacuum,' demonstrating a need to propose new legislation to deal with the changing issue of fetal rights.[73] In most jurisdictions, the fetus is not considered to have a 'legal personality' until it is born alive, at which point it is granted full human status.[74] A number of potential issues are immediately apparent when considering ectogenesis in this context. As a fetus incubated entirely artificially would not be 'born' in the traditional sense, this legal milestone would never be reached. For a fetus transferred from its mother's womb into an artificial womb, legal personhood would be temporarily bestowed during transit and then revoked on arrival according to this logic. This demonstrates that merely establishing rights for ectogenesis fetuses as equal to those for naturally gestated fetuses would not be sufficient, and that a number of specific laws would have to be written to deal with this unique situation. Such an activity would first require defending the artificial womb as a non-identical, but nevertheless non-inferior, gestational environment. Much like in natural pregnancies, it is likely that the viability criterion would also play a role in deciding when an ectogenesis experiment or 'pregnancy' could be terminated, as there is legal precedent for considering life-sustaining technologies as existing beyond the legally mandated standard of care every citizen has the right to receive. Just as life-support systems can be switched off when their continuation is deemed to be against the best interests of the patient, so could an artificial gestation be terminated on the discovery of a serious, debilitating condition. However, it is unlikely a termination could be justified *without* medical grounds, since any interest the state, or indeed the fetus itself, may have in preserving fetal life would no longer have to be weighed against the mother's bodily integrity. Maintaining the significance of viability could serve to alleviate some of the concerns regarding using unproven ectogenesis technology, as before the point of viability the experiments could be discontinued if they appeared detrimental, whereas after the point of independent viability standards of care currently promote artificial incubation in a humidicrib so the situation is functionally equivalent.

Regarding the need to defend the non-inferiority of artificial gestation, there are other concerns that ectogenesis could lead to further commodification of children and the dehumanisation of the reproductive process. However, while commercial adoptions and other ARTs remain permissible, I argue these concerns do not justify prohibiting ectogenesis.[75] Smajdor also disregards concerns that ectogenesis babies will be deprived of an essential 'bonding' experience with their gestational mothers, by noting the 'huge disservice' such opinions do to all adoptive and step-parents who harbour great affection for their children, not to mention all fathers who, due to biological restrictions, are unable to 'bond' through gestating their offspring.[76] The advent of ectogenesis is likely to lead to a radical perceptual shift regarding the norms of parenthood, providing an opportunity to challenge certain entrenched values. Bayne and Kolers claim there are currently 'three distinct views' about what makes one a parent: '*geneticists* claim that parenthood arises from direct genetic derivation; *gestationalists* claim that parenthood arises from gestation and birth; and *intentionalists* claim that parenthood arises from intentions to create, nurture and rear.'[77] They note that those who ascribe to the gestationalist account of parenthood not only disqualify all men from becoming parents, but also imply that 'infants gestated in an artificial womb would be orphans.'[78] While gestationalists defend this view by appealing to notions of infant attachment and maternal bonding, Bayne and Kolers note there is 'little evidence' that the fetus becomes attached to its gestational mother 'either in the womb or the birth process.'[79] Furthermore, they note that the emotional strain of childbirth often means new mothers feel 'indifferent to their new-borns' for at least a week after birth.[80] Thus, it appears that removing the necessity of gestation and childbirth from procreation could benefit some mothers while having, at worst, a negligible impact on their offspring. Most importantly, ectogenesis could challenge certain myths regarding maternal bonding that are often used to subjugate women and restrict their behaviour while pregnant and caring for young children.

There is one way though in which babies created through ectogenesis could be considered disadvantaged compared with physically gestated babies – they may be more likely to be denied the benefits of breastfeeding. As the process of pregnancy is responsible for initiating lactation, the removal of this process would usually carry with it the necessity of formula feeding for ectogenesis babies. It is accepted wisdom that breast milk is the ideal source of nutrition for infants, and Sander-Staudt notes scientists have 'not been able to identify, much less duplicate' all the components of breast milk responsible for providing nutrition and passive immunity transfer.[81] As such, being deprived of this source by default could be considered an unfair disadvantage for babies created through this technology, resulting in a failure of the Pareto test. However, not all biological mothers are able to breastfeed and many choose not to for various reasons, leaving bottle-fed ectogenesis babies no worse off by comparison. Furthermore, there have been cases where lactation has been successfully induced in the commissioning mother of a surrogacy arrangement, demonstrating that hormonal stimulation could be used for mothers who desire artificial gestation but natural lactation.[82] Although babies born through ectogenesis may be more likely to be bottle-fed for simplicity's sake, this does not leave them at any greater disadvantage than babies born naturally whose mothers are either unable, or simply choose not to breastfeed.

Overall, while the need to protect the interests of the fetus requires careful development and implementation of ectogenesis technology, I argue the challenges surrounding the contentious issue of fetal rights do not constitute an insurmountable obstacle to the ethical provision of this service.

Protecting against disability discrimination

When describing the potential benefits of ectogenesis, Sander-Staudt claims this technology could 'prevent the tremendous suffering of women as a result of the loss of a wanted pregnancy...miscarriages would be almost unheard of in the case of a chromosomally healthy fetus.'[83] What I believe is exposed in such a statement though is that there may be grounds to consider ectogenesis a potential threat to disability rights, particularly as many bioethicists believe the technology will be inevitably accompanied by pre-implantation genetic diagnosis (PGD) and subsequent selection against disabled embryos. There are even some who believe such ARTs will lead to genetic engineering and the 'social acceptance of eugenics.'[84] As such, I argue it is necessary to address the issue of disability discrimination to ensure ectogenesis is not used to diminish opportunity for persons with disabilities.

In the 1970s, when debates regarding the legalisation of abortion came to the fore, one of the major justifications used to support legal access was aimed at preventing the birth of children with severe disabilities. C. Cameron and R. Williamson note that even in countries where abortion is not available on demand, there are often special allowances for terminating pregnancies when it is known the fetus is damaged in some significant way.[85] Disability rights activists often challenge the validity of this distinction, claiming that selective abortion of disabled fetuses constitutes discrimination against people currently living with disabilities in society, by engendering an attitude in which the lives of disabled individuals are considered less valuable than those of able-bodied individuals. Furthermore, disability rights lawyer, Harriet McBryde Johnson, posits that the underlying assumptions regarding the quality of life of persons living with disabilities are often fundamentally flawed, leading to a perceptual exaggeration of the disadvantages these people face.[86]

Australian disability advocates Helen Houghton and Christopher Newell write: 'If prenatal diagnosis is used to eradicate as many disabilities as possible in society, then it does discriminate against people with disabilities.'[87] One of the major reasons given in support of this argument rests on the assumption that if the number of disabled people decreases, funding and support for those remaining will become eroded. According to Lynn Gillam, various studies have shown that following a positive prenatal test for conditions like spina bifida, Tay-Sachs, Down's syndrome, anencephaly or thalassemia, the vast majority of women will choose to terminate their pregnancies.[88] As PGD does not require selective abortion it is likely that even more women and couples will select against the implantation of an embryo known to have a serious defect.[89] As such, both forms of prenatal testing could result in a significant reduction in the births of people with disabling conditions, thus diminishing their numbers in society. However, Gillam notes that the belief that disability services would therefore become eroded neglects the possibility that with a reduction in the total number of disabled people those remaining would have greater access to services, as they would be in less demand.[90] It is also relevant to consider that a lot of the measures required to accommodate persons with disabilites, such as installing ramp access to public buildings or providing large-print versions of reading materials, are also necessary to serve the needs of the elderly, and thus would not be impacted by a reduction in the number of disabled users.

Jeff McMahan identifies another common objection to selective abortion on the grounds of disability: that such a deliberate attempt to reduce the number of disabled people coming into existence is morally equivalent to a racist society attempting to eradicate a racial minority.[91] Remembering that in any society with a liberal abortion law women are permitted to terminate a pregnancy for *any* reason, I argue that consistency dictates that they be allowed to select against disability. Furthermore, Cameron and Williamson remind us that most disabilities are not genetic, but rather are acquired as a result of accidents, which society actively tries to prevent.[92] They note it would not be plausible to accuse a traffic safety campaign of discriminating against the disabled, even though the intended result may be to reduce the number of disabling injuries acquired due to traffic accidents. John Harris also makes the valid point that *even if* rejection of a disabled embryo or fetus could be held to be discriminatory, it is not a disabled *person* who is being discriminated against, but rather an entity generally considered to be of much lower moral status. This is evident by the fact the fetus or embryo does not have any inherent claim to existence, and thus cannot be said to be unfairly disadvantaged due to parental preference for an able-bodied child.[93] While PGD and selection against disabled embryos may not be justifiably labelled discrimination against people living with disabilities, Gillam notes there is still the potential for such practices to foster discriminatory *attitudes* toward the disabled, which would need to be actively opposed in order to avoid any loss of opportunity within this group.[94] Disability anti-discrimination laws and state-funded disability services must therefore remain legally protected, in order to avoid undue disadvantage if PGD and other related technologies prove to have a significant impact on the number of people with disabilities living within the community.

There are also concerns that with increased testing available, parents who choose not to test for or select against disability may come to be considered morally or legally culpable for this decision.[95] Rather than disability being treated as a misfortune that the state has some responsibility to compensate for, this perceptual shift carries the risk of attributing the birth of a disabled infant, either naturally or via ectogenesis, to the 'conscious choice' of the parents, thereby justifying imposing the entirety of the financial burden of raising a special needs child on the individual family involved.[96] In his famous article 'Procreative Beneficence,' Julian Savulescu claims that parents have a *moral obligation* to create the 'best children' possible, including through prenatal testing, particularly if the tests impose minimal costs on the couple involved.[97] As PGD imposes fewer physical risks on the mother than other forms of prebirth testing, the pressure to utilise this service is already greater for IVF users than other women, and could possibly be greater still for ectogenesis clients, demonstrating a need to promote autonomous decision making among these groups. Furthermore, the right to refuse prenatal testing must be respected, regardless of whether the pregnancy in question is the result of natural conception, IVF or artificial gestation. This arises from the need to respect patients' wishes, including any desire *not* to know their own health status or that of their unborn children. As we do not currently force pregnant women to undergo prenatal testing, it would be unethical to impose this on women seeking to use ARTs, despite the fact these tests would be less invasive in such cases. Nevertheless, it is likely many couples who choose to avail themselves of new ARTs will also take the opportunity to screen their embryos for genetic abnormalities.

According to Gillam, selective abortion on the grounds of disability is often considered 'deeply offensive and hurtful' by persons living with disabilities, often due to the interpretation of this action as belying a quality of life assessment that claims a child is 'better off dead than disabled.'[98] However, as no fetus has the right to *demand* it be brought into existence, and abortion is permitted as a solution to unwanted pregnancies in general, I argue that, on the grounds of consistency, selective abortion must be permissible as a solution to a specific unwanted pregnancy, regardless of the reasons behind this. June Mary Zekan Makdisi notes that decisions regarding PGD are often easier on prospective parents than prebirth testing, precisely because these decisions are 'relating not to the unborn, but to the unimplanted.'[99] Some theorists therefore use the term 'pre-embryo' to refer to any zygote awaiting implantation, suggesting this entity should have a lower moral status than an embryo or fetus that is already partially developed *in utero*.[100] Indeed, one of the major drawbacks of current prenatal tests is that they can be conducted only when the fetus has already developed to a particular stage.[101] Therefore, were ectogenesis to lead to an increase in the uptake of PGD, this could reasonably be expected to lead to a decrease in selective abortions of disabled fetuses.

PGD before ectogenesis or physical pregnancy could also relieve a lot of the psychological stress of what Rothman calls the 'tentative pregnancy,' in which pregnant women are encouraged to await confirmation that their fetus is 'undamaged' before fully embracing their pregnant state.[102] This is a recent phenomenon affecting women as a direct result of the increase in prenatal tests available and the medicalisation of pregnancy. According to Rothman, women who are initially socially pressured to alter their lifestyle for the benefit of their 'baby' are, following an unhappy prenatal test such as an amniocentesis, encouraged to view the same entity as a collection of defective cells that should be aborted. PGD allows these tests to be carried out *before* the woman has had to gestate the fetus for any period of time, thereby eliminating the need for women to endure the burdens of the first trimester and beyond before discovering they do not wish to continue their pregnancy. In the case of ectogenesis, PGD could prevent the implantation of a disabled embryo that the parents might later wish to abort. As Smajdor observes, most wanted pregnancies now are more accurately 'conditionally wanted,' with the condition being that the resultant offspring not be affected by any debilitating condition.[103] Thus, PGD could enhance choice and informed consent regarding pregnancy, for those intending to physically or artificially gestate. As the majority of prebirth tests that indicate disability are followed by abortion, even if ectogenesis did increase the likelihood that PGD would be used to select against a disabled embryo, this embryo would be no worse off than were it allowed to continue to develop only to be aborted later. The termination of a pregnancy mid-gestation following later discovery of a defect could also be less traumatic in the case of ectogenesis, as women would not have to undergo the invasive procedures necessary to procure an abortion and would not have been gestating the fetus within their own bodies. So while disability rights should be actively promoted and biased quality of life evaluations for persons living with disabilities be abandoned in society, I argue the concerns raised by disability rights activists do not require the prohibition of ARTs, including ectogenesis. Whether or not prospective parents should be allowed to use future technologies to select *for* desirable traits or enhancements would be the subject for a whole other book. Suffice it to say for the purposes of this argument that ectogenesis would not create any *unique* problems in this area, as genetic manipulation of the embryo is already possible given existing IVF technologies and other interventions.

In summary for this chapter, by ensuring that no women or fetuses are unduly disadvantaged, I argue it is possible to protect equal opportunity while still pursuing the goal of ectogenesis. The use of Pareto optimality theory is particularly helpful in ensuring that providing more choice for some individuals does not serve to reduce options for others. By protecting the right to physical pregnancy and ensuring current anti-discrimination laws are rigorously enforced, in addition to respecting women's right to refuse technological intervention in the childbearing process, women who do not wish to use new reproductive technologies would be in no way harmed by the advent of ectogenesis services. Furthermore, by protecting legalised abortion in the interests of women's health and safety and promoting the interests of the fetus in ways that do not impose upon the rights of the mother, ectogenesis can be seen to provide welcome alternatives to unwanted pregnancies for some women, without diminishing options for others. In the final chapter of this book, I will explore how access to ectogenesis services could be provided for women who would seek them, focusing specifically on the Australian healthcare context.

Notes

1. Johnstone, 'Ethics and Ectogenesis,' 33; Edwards, 'New Conceptions,' 352.
2. Gerald Dworkin, 'Is More Choice Better than Less?' in *Ethical Principles for Social Policy*, ed. John Howie (Carbondale: South Illinois University Press, 1983), 79.
3. Nina G. Golden, 'Pregnancy and Maternity Leave: Taking Baby Steps towards Effective Policies,' *Journal of Law and Family Studies* 8 (2008): 1.
4. Paula McDonald, et al., 'Expecting the Worst: Circumstances Surrounding Pregnancy Discrimination at Work and Progress to Formal Redress,' *Industrial Relations Journal* 39, no. 3 (2008): 229.
5. Ibid., 242.
6. Heather Boushey, 'The Role of Government in Work-Family Conflict,' *The Future of Children* 21, no. 2 (2011): 170.
7. Golden, 'Pregnancy and Maternity Leave,' 9.
8. Ibid., 14; Erin M. Grabe, 'Gradual Return to Work: Maximizing Benefits to Corporations and Their Caregiver Employees,' *The Journal of Corporation Law* 37, no. 3 (2012): 699.
9. Golden, 'Pregnancy and Maternity Leave,' 22; 37.
10. Grabe, 'Gradual Return to Work,' 699.
11. Nancy Casas, 'Sex Discrimination on the Basis of Pregnancy: Australia's Report on Pregnancy Discrimination Should Make the United States Re-evaluate the Effectiveness of the Pregnancy Discrimination Act in Eliminating Pregnancy Discrimination in the Workplace,' *Transnational Law and Contemporary Problems* 11 (2001): 142.
12. McDonald, et al., 'Expecting the Worst,' 237.
13. HREOC, 'Pregnant and Productive: It's a Right Not a Privilege to Work While Pregnant. Report of the National Pregnancy and Work Inquiry,' (1999): xxi.
14. Casas, 'Sex Discrimination on the Basis of Pregnancy,' 146; 167.
15. McDonald, et al., 'Expecting the Worst,' 244.
16. Lenore Taylor, 'Human Rights Commission President Gillian Triggs Hits Back at Critics,' *The Guardian*, 1 April 2015. Available at: http://www.

theguardian.com/australia-news/2015/apr/01/human-rights-commission-president-gillian-triggs-hits-back-at-the-critics.
17 Jeanne Hayes, 'Female Infertility in the Workplace: Understanding the Scope of the Pregnancy Discrimination Act,' *Connecticut Law Review* 42, no. 4 (2010): 1307; 1310.
18 Donchin, 'The Future of Mothering,' 123.
19 Murphy, 'Is Pregnancy Necessary?' 75.
20 Fischer, et al., 'Pursuing Parenthood,' 428.
21 Lauritzen, *Pursuing Parenthood*, 32.
22 Cannold, 'Women, Ectogenesis, and Ethical Theory,' 55.
23 Murphy, 'Is Pregnancy Necessary?' 72.
24 Fischer, et al., 'Pursuing Parenthood,' 429.
25 Lauritzen, *Pursuing Parenthood*, 39.
26 There is particular concern regarding the psychological well-being of women and couples undergoing fertility treatment (Susan Caruso Klock and Dorothy A. Greenfield, 'Psychological Status of In Vitro Fertilization Patients during Pregnancy: A Longitudinal Study,' *Fertility and Sterility* 73, no. 6 [2000]: 1159–64).
27 Lauritzen, *Pursuing Parenthood*, 31.
28 Martha E. Gimenez, 'The Mode of Reproduction: A Marxist-Feminist Analysis of the Effects of Reproductive Technologies,' *Gender and Society* 5, no. 3 (1991): 337.
29 Lauritzen, *Pursuing Parenthood*, 32.
30 Singer and Wells, 'Ectogenesis,' 19.
31 Rosalind Ekman Ladd, 'Women in Labor: Some Issues about Informed Consent,' in *Feminist Perspectives in Medical Ethics*, eds Helen Bequaert Holmes and Laura M. Purdy (Bloomington: Indiana University Press, 1992), 217.
32 S. Oliver, et al., 'Informed Choice for Users of Health Services: Views on Ultrasonography Leaflets of Women in Early Pregnancy, Midwives, and Ultrasonographers,' *British Medical Journal* 313, no. 7067 (1996): 1252.
33 Richard Johanson, et al., 'Has the Medicalization of Childbirth Gone Too Far?' *British Medical Journal* 324, no. 7342 (2002): 892.
34 Gimenev, 'The Mode of Reproduction,' 334.
35 Elizabeth Heitman, 'Social and Ethical Implications of *In Vitro* Fertilization,' *International Journal of Technology Assessment in Health Care* 15, no. 1 (1999): 24.
36 Gimenev, 'The Mode of Reproduction,' 337.
37 Lockwood claims this is true even for those patients for whom such technology may represent substantial risk, such as IVF referrals for transplant recipients, HIV-positive women and multiple sclerosis sufferers (Lockwood, 'Pregnancy, Autonomy and Paternalism,' 538).
38 Fischer, et al., 'Pursuing Parenthood,' 431.
39 Lori B. Andrews, 'My Body, My Property,' *The Hastings Center Report* 16, no. 5 (1986): 34.

40 Heriot, 'Fetal Rights and the Female Body,' 183.
41 Nombuso Shabalala, 'A New World Court to Judge Gender-Based War Crimes,' *Agenda: Empowering Women for Gender Equity* 52 (2002): 89.
42 Steven L. Ross, 'Abortion and the Death of the Fetus,' *Philosophy and Public Affairs* 11, no. 3 (1982): 234.
43 Cannold, 'Women, Ectogenesis, and Ethical Theory,' 55.
44 Ibid., 57.
45 Ross, 'Abortion and the Death of the Fetus,' 233–4.
46 Steiger, 'Not of Woman Born,' 146.
47 Barbara Hewson, 'Reproductive Autonomy and the Ethics of Abortion,' *Journal of Medical Ethics* 27, sup. 2 (2001): ii12.
48 Doherty, 'Could We Care for Amillia,' 768.
49 Alghrani and Brazier, 'What Is It?' 73.
50 Ross, 'Abortion and the Death of the Fetus,' 238.
51 Cannold, 'Women, Ectogenesis, and Ethical Theory,' 59, 61.
52 Ross, 'Abortion and the Death of the Fetus,' 249.
53 Alghrani and Brazier, 'What Is It?' 60.
54 Colker, *Pregnant Men*, 145.
55 Edwards, 'New Conceptions,' 352.
56 Ross, 'Abortion and the Death of the Fetus,' 236.
57 Ibid., 240.
58 Nicky Woolf, 'Purvi Patel Found Guilty of Feticide and Child Neglect over Unborn Baby's Death,' *The Guardian*, 5 February 2015. Available at: http://www.theguardian.com/us-news/2015/feb/04/purvi-patel-found-guilty-feticide-unborn-childs-death.
59 A. Biscoe and G. Kidson-Gerber, '"Avoidable" Death of a Jehovah's Witness with Acute Promyelocytic Leukaemia: Ethical Considerations and the Internal Conflicts and Challenges Encountered by Practitioners,' *Internal Medicine Journal* 45, no. 4 (2015): 461–2.
60 Amy Corderoy, 'Pregnant Jehovah's Witness' Decision to Refuse Treatment "Harrowing" for Hospital Staff after Mother and Baby Die,' *The Sydney Morning Herald*, 6 April 2015. Available at: http://www.smh.com.au/nsw/pregnant-jehovahs-witness-decision-to-refuse-treatment-harrowing-for-hospital-staff-after-mother-and-baby-die-20150406–1mf570.html.
61 D. Isaacs, 'Moral Status of the Fetus: Fetal Rights of Maternal Autonomy?' *Journal of Paediatrics and Child Health* 39, no. 1 (2003): 58–9.
62 Helena Anolak, 'Our Bodies, Our Choices: Australian Law on Foetal Personhood,' *Women and Birth* 28, no. 1 (2015): 60–4.
63 Murphy, 'Is Pregnancy Necessary?' 71.
64 Ibid., 72.

65 Alghrani and Brazier, 'What Is It?' 70.
66 H.L. Halliday, 'Surfactants: Past, Present and Future,' *Journal of Perinatology* 28, sup. 1 (2008): S55.
67 Alghrani and Brazier, 'What Is It?' 76–7.
68 Singer and Wells, 'Ectogenesis,' 22.
69 Peter Singer, 'Technology and Procreation: How Far Should We Go?' *Technology Review* 88 (1985): 22–30.
70 Jones, *Brave New People*, 111.
71 Nozomi Ishii, et al., 'Outcomes of Infants Born at 22 and 23 Weeks' Gestation,' *Pediatrics* 132, no, 1 (2013): 62–71.
72 Alghrani and Brazier, 'What Is It?' 68.
73 Edwards, 'New Conceptions,' 358.
74 Alghrani and Brazier, 'What Is It?' 52.
75 While Australia does not permit commercial adoptions *per se*, significant 'processing' fees to international adoption brokers are tolerated. Some of these fees are so substantial as to make the process arguably equivalent to commercial adoption.
76 Smajdor, 'The Moral Imperative for Ectogenesis,' 342.
77 Bayne and Kolers, 'Toward a Pluralist Account of Parenthood,' 221.
78 Ibid., 223.
79 Ibid., 233.
80 Ibid.
81 Sander-Staudt, 'Of Machine Born?' 118.
82 F.P. Biervliet, et al., 'Induction of Lactation in the Intended Mother of a Surrogate Pregnancy: Case Report,' *Human Reproduction* 16, no. 3 (2000): 581.
83 Sander-Staudt, 'Of Machine Born?' 113.
84 Rowland, 'Technology and Motherhood,' 513; June Mary Zekan Makdisi, 'Genetically Correct: The Political Use of Reproductive Technology,' *Pepperdine Law Review* 32, no. 1 (2004–5): 30.
85 C. Cameron and R. Williamson, 'Is There an Ethical Difference between Preimplantation Genetic Diagnosis and Abortion?' *Journal of Medical Ethics* 29, no. 2 (2003): 90–1.
86 Harriet McBryde Johnson, 'Unspeakable Conversations or How I Spent One Day as a Token Cripple at Princeton University,' *The New York Times Magazine*, 16 February 2003, 50.
87 Lynn Gillam, 'Prenatal Diagnosis and Discrimination against the Disabled,' *Journal of Medical Ethics* 25, no. 2 (1999): 163–4.
88 Ibid.; The results showed values between 70 and 100 per cent.
89 In Australia it is actually an offence to knowingly implant an embryo deemed 'unfit for implantation,' as defined by the *Ethical Guidelines on the Use of Assisted Reproductive Technology in Clinical Practice and Research* (2004). This includes embryos found to have a genetic defect through PGD.

90 Gillam, 'Prenatal Diagnosis,' 164.
91 Jeff McMahan, 'Causing Disabled People to Exist and Causing People to Be Disabled,' *Ethics* 116, no. 1 (2005): 82.
92 Cameron and Williamson, 'Is There an Ethical Difference,' 91.
93 John Harris, 'One Principle and Three Fallacies of Disability Studies,' *Journal of Medical Ethics* 27, no. 6 (2001): 386.
94 Gillam, 'Prenatal Diagnosis,' 163.
95 Joseph Losco and Mark Shublak, 'Paternal-Fetal Conflict: An Examination of Paternal Responsibilities to the Fetus,' *Politics and the Life Sciences* 13, no. 1 (1994): 63.
96 Gillam, 'Prenatal Diagnosis,' 166.
97 Julian Savulescu, 'Procreative Beneficence: Why We Should Select the Best Children,' *Bioethics* 15, no. 5–6 (2001): 413.
98 Gillam, 'Prenatal Diagnosis,' 170.
99 Makdisi, 'Genetically Correct,' 28.
100 Ibid., 4.
101 Alexander D. Wolfe, 'Wrongful Selection: Assisted Reproductive Technologies, Intentional Diminishment, and the Procreative Right,' *Thomas M. Cooley Law Review* 25, no. 3 (2008): 479–80.
102 Barbara Katz Rothman, *The Tentative Pregnancy: How Amniocentesis Changes the Experience of Motherhood* (New York: W. W. Norton & Company, 1993), 86.
103 Smajdor, 'In Defense of Ectogenesis,' 95.

3
Providing Equal Opportunity to Ectogenesis

Abstract: *Kendal considers a number of possible distribution methods for ectogenesis services, concluding that state sponsored services are an ethical requirement to ensure equality of opportunity is not compromised for women of low socio-economic status. Alternative allocation methods considered include a user-pay system, priority on the basis of medical need and utilitarian considerations. This discussion is set within Australia's public healthcare system, Medicare.*

Keywords: health equity; Medicare; public healthcare; redistributive justice

Kendal, Evie. *Equal Opportunity and the Case for State Sponsored Ectogenesis*. Basingstoke: Palgrave Macmillan, 2015. DOI: 10.1057/9781137549877.0007.

In this final chapter I will explore the issues surrounding the practical provision of ectogenesis services, should they become available in the future. In particular, I will discuss the ethical considerations for equitable distribution of these services, using the Australian healthcare system as a case study. It will be my contention that ectogenesis services should be state sponsored, and not made available privately until a robust public system is established that provides for the needs of all women, regardless of fertility status or financial situation. While it is reasonable to expect a private industry to arise from this technology, I argue that since ectogenesis has the potential to yield significant medical and social benefits to all women, access should not be limited to the wealthy alone. Thus, the ideals of equal opportunity are important in determining whether ectogenesis should be introduced into the range of ARTs currently eligible for subsidy through Australia's public healthcare scheme, Medicare. While I leave it to the economists to determine exactly how far such a subsidy might reasonably extend, I argue against any level of co-payment that would constitute an undue impediment to equal access to this technology for those of lower socio-economic status.

Australia's two-tier healthcare system

Although most industrialised nations had public healthcare systems in place by the early 1970s, Australia's Medicare program was launched only in 1984.[1] In spite of its late beginning, universal medical coverage was achieved within a few years, with all Australian citizens being entitled to treatment in public hospitals at no private expense.[2] Australia also boasts a significant private healthcare sector, with estimates in 2008 indicating approximately 32 per cent of total health expenditure arose in this sector, one of the 'highest proportions within the OECD.'[3] This unique blend of public and private expenditure has led to Australia's so-called two-tier healthcare system, in which the wealthy can purchase supplementary health insurance in addition to accessing standard public services.[4] In effect this means wealthier citizens can opt for private hospital care, shorter waiting lists and can often choose their own physicians, while public patients rely solely on publicly funded healthcare services. While this mixed system has received some criticism on the grounds of equity – a common criterion for evaluating the effectiveness of healthcare systems – Australia is often considered to have one of the more successful public schemes available for managing population health.[5]

According to Gwen Gray, there are various justifications for providing healthcare services from public money that rely on establishing health as a 'quasi-public good.'[6] The central argument supporting such expenditure is that universal healthcare benefits *all* members of a community, irrespective of who financially contributes to it. The example often used to illustrate this is infectious disease control, which has the potential to significantly reduce morbidity and mortality across an entire population. Other justifications include the need to support a healthy labour force to maximise productivity, or appeal to communitarian ideals of charity, shared responsibility, and the establishment of healthcare as a basic human right the state has a duty to provide.[7] Australia's healthcare system appears to embody most of these ideals, adhering to a 'principle of universality' that dictates the duties of a humane society to its sick and injured members are the same regardless of their socio-economic status.[8] Although Medicare is praised for its comprehensive coverage, there are various sub-groups within the Australian population that are known to experience significantly lower health status compared to the general population, most notably including rural and Indigenous Australians.[9] There are also concerns that universal access to healthcare in Australia is being eroded, partially as a result of the private sector's relationship with the public system. The term 'parasitic' has been used to describe the manner in which private patients still use public services for most major surgeries, as well as how significant public education funding is often used to train doctors who then go into private practice.[10] Thus, the most disadvantaged members of the population are failing to receive the same standard of care as wealthier citizens who have access to both public and private healthcare resources. From both an equal opportunity and feminist perspective these issues are major ethical concerns, particularly as women within these vulnerable populations often represent an even further disadvantaged group. Geographical isolation is also linked to greater resource rationing, further diminishing the quality of public healthcare available in remote areas.

The significant disparity between the wealthier and poorer members of the population, and between urban and rural patients, is particularly evident in the case of access to ARTs.[11] While many ethicists argue that reproductive rights should be asserted as positive rights that the state has a duty to support, the allocation of scarce healthcare resources to these technologies remains contentious.[12] I propose that Australia's healthcare system provides an ideal setting in which to explore the potential for both publicly and privately funded ectogenesis services, particularly as Medicare already subsidises various infertility treatments for public patients. Even though infertility is not a life-threatening condition, research indicates it significantly impacts people's quality of life, thus making it a public health issue.[13] Throughout the rest of this section I will explore various methods for healthcare rationing as might apply to the future provision of ectogenesis in Australia, arguing that developing a state sponsored system is vital for the ethical distribution of these services.

Providing access regardless of ability to pay

A common ethical concern with regard to surrogacy, and one that is famously dramatised in Margaret Atwood's feminist dystopia *The Handmaid's Tale* (1985), is the possibility that such a development would lead to the formation of an underclass of women whose reproductive labour is hired out to wealthier citizens. According to Murphy, a similar class system could arise from ectogenesis, with the social elite paying for these technologies, while poorer communities still rely on women's bodies for gestation. She relates a story in which a researcher was willing to petition the courts for permission to sustain a pregnancy in the uterus of a brain-dead patient, on the grounds that women's bodies are still 'the cheapest incubators we have.'[14] While such cases are still rare, they are unfortunately being reported with increasing frequency in medical literature, highlighting the fact that artificial incubation is still considered a far more expensive alternative to physical gestation, even if the mother's body requires total life support.[15] However, given the concerns Murphy raises about ectogenesis, I argue it is not acceptable to provide such a service only to those able to privately afford it. From a feminist perspective it is also important to consider how equality with men, and between women, would be affected if ectogenesis were restricted to a particular subset of women. As IVF access and adoption protocols have already shown, such restrictions, when permitted, often follow a classist and even racist agenda, preferentially accommodating the needs of affluent, white, heterosexual couples, while ignoring the needs of homosexual individuals, single women, minorities and people of low socio-economic status.[16]

David Lairson et al. note that the Australian healthcare system is already criticised for favouring higher-income individuals, as they are the ones 'most likely to carry both basic and supplemental hospital insurance.'[17] This is despite the fact that people living in rural areas and those of lower socio-economic status are known to be in greater need of healthcare resources.[18] In terms of reproductive health and fetal wellbeing, these sub-groups also experience poorer outcomes compared with higher-income citizens, partially due to the increased rates of smoking and alcohol consumption during pregnancy, poor maternal nutrition, and restricted access to prenatal care that correlates with lower socio-economic status and rural disadvantage.[19] Among Australia's Indigenous population infant mortality is approximately seven times the national average, with overall life expectancies also reduced by up to 20 years compared to non-Indigenous Australians.[20] In fact, the health status of Australia's Indigenous population is often considered comparable to that found in many developing countries. As poverty and rurality are major contributing factors to this disparity, increased public health funding to these communities is required to address this inequity.[21] With regard to ectogenesis, the increased risks associated with pregnancy and childbirth in remote areas with limited maternity care suggests that rural Australians may have a higher claim to these services than the (usually wealthier) urban population. This is particularly true when considering that ectogenesis would make it possible to have more equitable distribution of high technology care, as it provides a unique situation in which the mother and fetus need not be in the same geographical location.[22]

When treating ectogenesis as a medical technology aimed at preventing the health risks associated with pregnancy and childbirth, it is clear that allowing a user-pay system of distribution would serve to further disadvantage certain women. While libertarian philosophers like Robert Nozick would suggest that individuals should be free to enjoy whatever 'natural advantages' they possess, including better access to healthcare and medical technology due to higher income, I argue that if wealthy women were the only ones able to afford ectogenesis, this would make them even more competitive in areas like employment, where they are already at an advantage.[23] This is in direct conflict with the ideals of equal opportunity and the principle of universality on which Medicare was founded. According to Uwe Reinhardt, Nozick's libertarian vision for a free market healthcare system is not shared by 'the world at large,' as 'literally no country seems prepared to surrender the delivery of personal health services and products to arbitration by unfettered market forces.'[24] In Australia, not even the private healthcare sector is abandoned to the mercy of the free market, with various government mandates dictating the minimum coverage standards private insurance providers can supply. Furthermore, access to public healthcare resources cannot be refused to privately insured citizens.[25] As wealthier citizens currently enjoy greater healthcare benefits than poorer citizens, I argue that allowing artificial gestation to become a luxury product would further compound the inequalities that already lead to health disparities between socio-economic groups in society.

Establishing that ectogenesis should not be available only to the wealthy does not negate the possibility of a substantial private market developing, as has been the case for other ARTs. In much the same way as Australia's two-tier healthcare system currently allows for both public and private hospital coverage, I argue ectogenesis services could be provided within both sectors without significantly disadvantaging citizens of lower socio-economic status. Unlike in the case of primary health services, allowing wealthier citizens the option of paying a premium for private ectogenesis programs, while public patients experience longer waiting periods to access state sponsored services, is unlikely to cause substantial harm to poorer citizens. After all, one does not need 'emergency' reproduction, the way one may need emergency surgery or other potentially life-saving medical procedures. It is for this reason that I believe Australia's two-tier healthcare system makes it an ideal candidate to pioneer the ethical provision of future ectogenesis technology. With well-established public and private sectors for ARTs, Australia could serve as an example to other countries regarding how these sectors could work together to provide equitable access to ectogenesis, while avoiding the class system that Murphy fears. As long as a robust public system is available from the outset, allowing private industry to develop in this way would not disadvantage poorer citizens who could not afford user-pay access, and would likely reduce financial strain on the Medicare system.

Providing access according to need

Having already dismissed a wholly user-pay system, I will now explore the argument for rationing access to ectogenesis on the grounds of medical necessity. In *Just Health: Meeting Health Needs Fairly*, Norman

Daniels argues that health should be placed in a special category of goods, and that the state has a responsibility to subsidise healthcare to protect each citizen's access to the 'normal opportunity range' for someone at their stage of life.[26] As health is a 'necessary condition' for access to numerous other social, educational and employment opportunities, many ethicists agree that the state should intervene in health service provision, rather than allow market values to determine health outcomes.[27] However, Daniels also argues that there is a limit to what is appropriate for the state to fund in terms of healthcare, claiming that the goal of public healthcare should only be to promote, restore or maintain 'normal species functioning.'[28] According to this theory, it is possible to justify public funding for ARTs only for physiologically infertile women of 'normal' reproductive age. While this form of rationing is less ethically compromising than allowing an open market to determine access to ectogenesis, I argue it fails to uphold the ideals of equal opportunity. As explored in the Introduction, there are medically reasonable grounds for *all* women to request access to a technology that could eliminate the physical burdens and risks of pregnancy and childbirth, especially when considering the additional social and economic burdens that accompany this form of reproduction. Furthermore, even in industrialised nations like Australia, pregnancy always carries with it some risk of severe injury, or even death. As such, I contend the usual criterion for public healthcare funding, namely, medical necessity, is satisfied in every case, as ectogenesis would eliminate not only the *risk* of serious complications, but also the *certainty* of pregnancy-related discomforts and the physical trauma of childbirth.

As demonstrated, there are reasons not to restrict public funding for ectogenesis to the physiologically infertile. However, I would also argue there are grounds to prevent harsh age restrictions on those accessing the technology. Again, the rationale for this originates both from a feminist and equal opportunity perspective. Remembering that men are usually able to father children at advanced ages, if the goal of ectogenesis was to bring men and women closer to equality in reproductive endeavours, providing access only to younger, pre-menopausal women would fail to achieve this. Furthermore, as fertility rates continue to decline in Australia among all age groups under 30, there is sufficient evidence to suggest women are opting for postponement of childbearing, most likely due to career commitments.[29] Hayes notes such a postponement may increase women's employment opportunities, but has a detrimental effect on their fertility as 'even at the relatively young age of thirty, up to ninety percent of a woman's eggs are gone.'[30] Brigitte Leeners et al. note that fertility steadily declines with age for women, with a reduction of 6 per cent for 25–29-year-olds, 14 per cent for 30–34-year-olds and 31 per cent for 35–39-year-olds, with much larger reductions seen after this.[31] Even using IVF, the chance of achieving a live birth in a woman over 42 years of age is only 4 per cent, compared with 41 per cent for a woman aged 35.[32] Owing to the limited window of opportunity they have to start a family, women are often forced to temporarily resign from paid employment at crucial times in their career development, whereas men generally are not. Thus, I argue a system that allows for postponed childbearing would enhance women's employment opportunities and increase equality, not only with men but also between women, as the age at which each individual woman finds a suitable partner with which to procreate can vary widely, thereby significantly impacting the likelihood of reproductive success. The benefits of achieving financial stability before starting a family are numerous; however, the risks to the fetus, particularly related to certain chromosomal abnormalities, increase with advanced maternal age. Assuming ova were harvested at a younger age, when both quantity and quality are better, ectogenesis could serve to satisfy the desires of many to delay starting a family, without the associated risks to fetal wellbeing. Even in cases where eggs from an older woman were used, these could be screened for abnormalities before proceeding. Thus, there is legitimate reason to support access to ectogenesis among older women, particularly as longer life expectancies mean offspring are unlikely to be orphaned as young children even if their parents choose to produce them in their fifth or sixth decade of life.[33]

There are those who would consider the use of ectogenesis by postmenopausal women to be an 'enhancement' rather than a medical 'treatment,' as it would extend women's reproductive lives beyond that which occurs in nature. For Daniels this would fall outside the boundary of maintaining 'normal species functioning,' and therefore be beyond what the state should provide.[34] However, as Smajdor notes, even 'natural inequalities,' such as the restrictive biological window for females to reproduce compared with males, are 'candidates for redistributive justice.'[35] In healthcare system evaluation the ideals of redistributive justice demand that any 'redistribution' of resources be done in such a manner as favours the most disadvantaged within the system.[36] I argue the concept of redistributive justice could therefore be used to defend access to publicly funded ectogenesis, regardless of maternal age, as the impact of age on fertility is unequally distributed between the sexes. Rather than viewing this technology as an enhancement for older and even post-menopausal women, it could be seen merely as a method of addressing a naturally occurring biological inequality. However, current age restrictions on ARTs indicate that attitudes within society and the medical community would need to change before the use of ectogenesis for post-menopausal women could be achieved.

Defining 'need' in the way described previously would also serve to eliminate some of the discriminatory practices currently present in the provision of ARTs. When considering that the physiologically fertile do not have to submit to screening processes before being allowed to procreate, many consider it ethically problematic that IVF patients and prospective adoptive parents do.[37] By making access to the technology about the health of the woman, rather than social approbation of her potential parenting ability, the discriminatory practices of the past can be avoided. Where a history of adultery, same-sex attraction or poverty were once grounds for forced sterilisation and losing custody of existing children, it is important to ensure new technologies are not co-opted to promote a socially repressive agenda.[38] This is particularly evident when promoting the use of ectogenesis as a solution for gay couples wanting to procreate, as historically gay and lesbian couples have had difficulty accessing ARTs in most countries, particularly if unable to demonstrate physiological infertility as well as social.[39]

Providing access according to utilitarian principles

While ensuring a publicly funded option would avoid inequality on the basis of socio-economic status, there would still be the need to establish guidelines for creating waiting lists if public demand exceeded supply capacities or funding limits. When dealing with the issue of scarce healthcare resource allocation, utilitarian considerations are often introduced to determine which public patients should be prioritised for state funded treatment. To place this in context, while classical utilitarianism aims to maximise overall *happiness*, utilitarian methods of healthcare allocation aim to maximise *health*, generally at the population level.[40] Jennifer Ruger claims that utilitarianism 'arguably serves as the standard framework for health policy analysis,' as methods of allocating healthcare services often rest on a 'maximisation principle,' – the greatest good for the greatest number.[41] As demonstrated previously, the potential health benefits of ectogenesis justify extensive research into this possibility, particularly as such a large proportion of the total population are susceptible to the health risks associated with pregnancy. In addition, there are substantial benefits to the fetus that ectogenesis could confer, suggesting this technology could make a significant contribution to improving overall population health.

Although utilitarian considerations may be used to support public funding for ectogenesis, I argue that they should not be the *sole* method of allocating access to this technology in the public healthcare system, as this may unfairly disadvantage certain groups. According to Oommen C. Kurian, utilitarianism is a popular framework for dealing with the issue of scarce resource allocation for its 'capacity to put forth a hypothetically objective base' for deciding matters of moral significance, such as prioritising who should receive state funded medical services.[42] It does this through the development of quantitative measures, such as cost–utility analysis (CUA) and quality-adjusted life year (QALY) calculations, which attempt to establish utility scores for different health conditions that 'can compare on a single qualitative scale.'[43] While the issue of incommensurability means that different health states cannot be directly compared, particularly without reference to the different adaptation abilities of the individuals involved, CUA and QALY calculations can yield useful information regarding health preferences. Since procreation is often considered a high social priority, this would seem to justify using public funds to assist citizens in achieving this goal. What I would like to consider here though is what Ruger calls the *priorities problem*, which looks at how much weight the most disadvantaged members of society should receive when allocating scarce healthcare resources.[44] When adhering to Jeremy Bentham's famous motto, 'each to count for one, and none for more than one,' everyone in the public healthcare system would be given equal weight in resource allocation decisions, without reference to social position.[45] However, such an approach would fail to take into consideration the health disparities discussed earlier in this chapter between rural and urban populations in Australia. As such, promoting equal opportunity access to ectogenesis may require prioritising public service provision among poorer communities, particularly as citizens within these communities also tend to experience poorer health outcomes, including in pregnancy and childbirth, and are the least likely to hold supplementary health insurance. Although I argue it should not form the *only* basis for allocating state sponsored access to ectogenesis, those citizens for whom this technology would represent the only possible method of procreating may have a higher claim to public funds in order to achieve this. Thus, prioritisation according to the *severity* of medical need could provide assistance in forming waiting lists for public services, even though the physical burdens associated with pregnancy and childbirth mean that all women would have a justified claim to access according to medical 'need.' This method of allocation would also prioritise access to limited ectogenesis chambers for babies born prematurely, as this constitutes a medical emergency, over embryos that have not yet been implanted.

The impact of certain social determinants of health means that utilitarian considerations alone cannot serve as the basis for healthcare allocation, but must be taken in conjunction with other methods of promoting social justice. When considering what ARTs should be publicly subsidised, strict utilitarianism can also serve to discriminate against minorities. Sexual orientation and other lifestyle factors, for example, often impact how extensive, and thus expensive, the technological interference required to achieve reproductive success.[46] Relying solely on a maximisation principle would exclude certain individuals from receiving state funded treatment by promoting 'overall utility' while continuing to tolerate inequality between individuals.[47] This is what Ruger calls the *aggregation problem* for utilitarianism, in which the needs of minorities are sacrificed in order to benefit the majority. Allocating resources according to purely utilitarian considerations would prioritise lower-cost cases and those most likely to succeed, in order to increase the total number of successes.[48] If equal opportunity is used as the basis to defend the expense of developing ectogenesis technology, it is clear the dominant utilitarian methodology of public healthcare rationing will need to be re-examined.

Qualitative research in Australia has shown that citizens are willing to abandon the maximisation principle in order to avoid discriminating against the less fortunate, indicating that other ethical considerations are relevant when regulating access to state funded healthcare and reproductive services.[49] So while the dominant utilitarian methodology of scarce health resource allocation may favour state sponsored ectogenesis in general, due to its potential to positively impact a large proportion of the population, it may serve as an obstacle to equal opportunity if not tempered with other considerations aimed at promoting social justice.

The real cost–benefit analysis

Recognising that all women, and perhaps men, might have a justifiable claim to state sponsored access to ectogenesis does not necessarily make it financially viable. In fact, Mary Anne Warren claims that any 'massive commitment to ectogenesis would probably be ruinously expensive.'[50] As such, I propose that an in-depth cost–benefit analysis should be conducted, and that the release of any future ectogenesis technology in Australia be dependent on Medicare's ability to fund access for public clients. Although it is not my intention to provide a detailed economic argument in favour of ectogenesis, there are some potential financial advantages to promoting this technology that warrant examination, especially given the substantial cost that would be involved in implementing a publicly funded system.

According to Lauritzen, one of the major obstacles when attempting cost–benefit analyses of ARTs is that 'many of the costs and benefits are incommensurable.'[51] He notes, for example, that when couples consider the financial burden of pursuing IVF, they are weighing this pecuniary cost against the potential benefit of creating their own child, a goal that is unlikely to carry a specific monetary value. At a population level, the costs of providing state sponsored access to ectogenesis in Australia may also need to be weighed against the significant burden on public health attributed to our ageing population.[52] Since conflicting life goals are often responsible for delayed childbearing and subsequent loss of fertility, I argue that a solution like ectogenesis, particularly if not restricted to younger women, could serve to alleviate some of this burden. As such, the additional cost of providing ectogenesis for women either unable or unwilling to become pregnant would need to be balanced against the concomitant benefit of population growth, which is generally considered a requirement for maintaining the working population and tax base needed to sustain the public healthcare system.[53] This is most relevant for countries like Australia that are experiencing population decline and are already relying heavily on immigration to compensate for reduced fertility rates.[54]

It is also possible that ectogenesis would decrease health expenditure in other areas, especially when used as a substitute for high-risk pregnancies. The significant drain on public resources that often results from IVF-induced multiple pregnancies could also be reduced were implantation failure not a major obstacle to overcome.[55] The costs of providing routine ultrasounds and other prenatal services would be decreased for women not choosing to pursue physical pregnancy, in addition to removing the cost of hospitalisation for childbirth. As 99 per cent of Australian women currently give birth in hospitals, and the majority will be referred to ultrasonography services at least once during their pregnancy, these are substantial expenses. This is in addition to the cost of lost productivity for women who must take time away from work both to give birth and to be treated for pregnancy-related illness.[56] The risk of requiring an emergency caesarean or other major (and costly) medical interventions during parturition would also be eliminated for women choosing artificial gestation, as well as avoiding the possibility of major disabilities in the infant resulting from physical birth trauma. In terms of fetal health, the ability to eliminate teratogen exposure for babies produced using ectogenesis could also reduce long-term health costs. If PGD was also used, severe disability due to chromosomal abnormalities could be prevented in many cases. Owing to these health advantages, Pence argues that private health insurance companies will likely include ectogenesis within their maternity services benefit schemes.[57] In order to preserve equal opportunity, I argue state funds must therefore be directed to provide access to ectogenesis for those individuals who do not hold supplementary health insurance.

It is also relevant to consider that many of the potential costs of ectogenesis are already being borne by the public healthcare system to some extent for babies born prematurely. In 2012, in Australia 8.5 per cent of babies were born prematurely, with over 1,267 live births occurring between 20 and 27 weeks gestation, 2,231 between 28 and 31 weeks gestation, and 44,218 between 32 and 36 weeks gestation.[58] The cost of sustaining NICU equipment and staff to provide for the needs of these babies already figures in the millions, and it is predicted that these costs will continue to rise as babies are born alive at earlier gestational ages.[59] In Australia it is estimated that each severely premature baby born costs approximately $3,000 a day to keep alive in NICU.[60] As such, research into more effective and efficient artificial incubation systems is already required to reduce the death rate among these infants, as well as the cost of treatment.

In summary, while state sponsored ectogenesis is likely to be an expensive endeavour, the demands of equal opportunity dictate that it should not be available only to the wealthy. Because of the physical, social and economic burdens associated with pregnancy and childbirth that have been discussed in earlier chapters, there are grounds to consider ectogenesis medically justifiable for *all* women, particularly as a preventive health measure aimed at achieving greater equity in health status across the Australian population. There are also certain economic advantages that would need to be considered in any cost–benefit analysis regarding Medicare's ability to subsidise access to this technology, including reductions in health expenditure in other aspects of maternity care, as well as the overall benefit of counteracting population decline. Regardless of what this analysis yields, it is important to remember that just because addressing inequality may be very expensive, this does not ethically justify a failure to do so. Many pregnancy anti-discrimination laws already in effect are extremely costly to workplaces and the public healthcare system, but upholding the ideals of equal opportunity demands that such measures are taken to protect the rights and opportunities of women regarding their reproductive choices.

Obstacles to providing ectogenesis services

Apart from the expense of developing and using ectogenesis services, there are various other obstacles that would need to be overcome before this technology could be successfully implemented in the Australian healthcare system. These include abandoning various restrictions on embryo and fetal experimentation for reproductive purposes.[61] In light of the fact that fetuses are not generally granted full personhood or legal status until birth, prohibiting research on surplus embryos or recovered fetal tissue seems contradictory. If ectogenesis research is to progress, laws will need to change. Given the substantial benefits to equality and opportunity for women, I contend that attempts to prohibit or forestall ectogenesis research are ethically questionable, whether they take the form of legal restrictions or social disapproval.

Even if the laws currently obstructing ectogenesis research do change in the future, by far the most serious impediment to this research appears to be society's attitude toward women. In their study concerning social attitudes toward ectogenesis in Israel, Simonstein and Mashiach-Eizenberg discovered sexist beliefs about women's role in procreation were at the heart of opposition to this technology. When phrasing survey questions about the need to develop an artificial womb from the perspective of saving fetuses, this study showed overwhelmingly positive results. However, these authors noted that 'when the idea of easing women's "natural" roles in reproduction was at the center of statements, the [artificial womb] became unacceptable.'[62] This substantiates Firestone's claim that ARTs are often 'excused' on the grounds of promoting the interests of fetuses, but rejected when focused on improving situations for women.[63] When attempting to account for the increased negativity of women toward ectogenesis, compared with men, Simonstein and Mashiach-Eizenberg claim that Israeli girls are socialised from early childhood that they have a 'duty' to procreate and that 'infertility is a curse.'[64] Thus, the influence of pronatalist dogma is further hindering social support for research into ectogenesis technology. This also plays out in the negative representation of ectogenesis in popular media, which in turn influences public opinion toward the technology.[65] I argue that any suggestion that women wishing to pursue biological motherhood through ectogenesis somehow lack commitment or are simply 'too posh to push' fails to recognise that a desire and commitment to parenthood does not necessarily entail a desire or commitment to being pregnant. If it did, no father or adoptive parent could ever satisfy this criterion.

Given the significant inequalities that arise as a result of reproductive difference, I argue the state has a responsibility to promote all reasonable methods to improve equal opportunity for women. This includes funding improvements to maternity care, enforcing anti-discrimination laws, funding education programs that challenge gender inequality and pronatalism, promoting research into more equitable methods of reproducing and, finally, providing state funded access to these new methods when they become available.

Notes

1 Gwen Gray, 'Access to Medical Care under Strain: New Pressures in Canada and Australia,' *Journal of Health Politics, Policy and Law* 23, no. 6 (1998): 909.
2 David Lairson, et al., 'Equity of Health Care in Australia,' *Social Science Medicine* 41, no. 4 (1995): 475.
3 Eddy van Doorslaer, et al., 'Horizontal Inequities in Australia's Mixed Public/Private Health Care System,' *Health Policy* 86, no. 1 (2008): 98.
4 Gray, 'Access to Medical Care under Strain,' 911.
5 Lairson, et al., 'Equity of Health Care in Australia,' 475.

6   Gray, 'Access to Medical Care under Strain,' 909.
7   Ibid.
8   Stephen R. Leeder, 'Achieving Equity in the Australian Healthcare System,' *Medical Journal of Australia* 179, no. 9 (2003): 476.
9   Matthew Tonts and Ann-Claire Larsen, 'Rural Disadvantage in Australia: A Human Rights Perspective,' *Geography* 87, no. 2 (2002): 137; Nigel Rice and Peter C. Smith, 'Ethics and Geographical Equity in Health Care,' *Journal of Medical Ethics* 27, no. 4 (2001): 256–61.
10 Gray, 'Access to Health Care under Strain,' 905; 911.
11 Danielle Herbert, et al., 'Early Users of Fertility Treatment with Hormones and IVF: Women Who Live in Major Cities and Have Private Health Insurance,' *Australian and New Zealand Journal of Public Health* 34, no. 6 (2010): 629.
12 Robert Blank, 'Assisted Reproduction and Reproductive Rights: The Case of In Vitro Fertilization,' *Politics and the Life Sciences* 16, no. 2 (1997): 279.
13 Jacky Boivin, et al., 'The Fertility Quality of Life (FertiQoL) Tool: Development and General Psychometric Properties,' *Human Reproduction* 26, no. 8 (2011): 2084–91.
14 Murphy, 'Is Pregnancy Necessary?' 69.
15 João Souza, et al., 'The Prolongation of Somatic Support in a Pregnant Woman with Brain-Death: A Case Report,' *Reproductive Health* 3, no. 1 (2006): 3.
16 Murphy, 'Is Pregnancy Necessary?' 72; Lavender, 'Lesbian Feminist Activism in Australia,' *Off Our Backs* 36, no. 3 (2006): 74.
17 Lairson, et al., 'Equity of Health Care in Australia,' 476.
18 Tonts and Larsen, 'Rural Disadvantage in Australia,' 132.
19 Anindya Sen, 'Is Health Care a Luxury? New Evidence from OECD Data,' *International Journal of Health Care Finance and Economics* 5, no. 2 (2005): 159.
20 Barbara Henry, et al., 'Institutional Racism in Australian Healthcare: A Plea for Decency,' *Medical Journal of Australia* 180, no. 10 (2004): 517.
21 Michael Kidd, et al., 'Primary Health Care Reform: Equity Is the Key,' *Medical Journal of Australia* 189, no. 4 (2008): 221.
22 For Indigenous Australians it is possible the ectogenetic fetuses would need to be transported to traditional lands for 'birth' for cultural reasons.
23 Robert Nozick, *Anarchy, State, and Utopia* (New York: Basic Books, 1974), 233.
24 Uwe Reinhardt, 'Resource Allocation in Health Care: The Allocation of Lifestyles to Providers,' *The Milbank Quarterly* 65, no. 2 (1987): 168.
25 For Australia it has been argued that a healthcare system founded on 'communitarian solidarity' is more likely to be sensitive to the needs of women and the Indigenous community, compared with a system based on the individualism of neo-liberalism, which seeks to achieve economic liberation through privatisation and support for deregulated open markets (Henry, et al., 'Institutional Racism in Australian Healthcare,' 519).

26 Norman Daniels, *Just Health: Meeting Health Needs Fairly* (Cambridge: Cambridge University Press, 2008), 175.
27 J. Wilson, 'Not So Special after All? Daniels and the Social Determinants of Health,' *Journal of Medical Ethics* 35, no. 1 (2009): 2; Gray, 'Access to Medical Care under Strain,' 909.
28 Daniels, *Just Health*, 29.
29 Farhat Yusuf and Stefania Siedlecky, 'Female Sterilizing Operations in New South Wales: A Demographic Perspective,' *Journal of the Australian Population Association* 15, no. 1 (1998): 75.
30 Hayes, 'Female Infertility in the Workplace,' 1301.
31 Brigitte Leeners, et al., 'The Relevance of Age in Female Human Reproduction – Current Situation in Switzerland and Pathophysiological Background from a Comparative Perspective,' *General and Comparative Endocrinology* 188 (2013): 169.
32 K. Mac Dougall, et al., 'Age Shock: Misperceptions of the Impact of Age on Fertility before and after IVF in Women Who Conceived after Age 40,' *Human Reproduction* 28, no. 2 (2013): 351. These authors note that many women are not well informed of the true impact of maternal age on fertility, with many assuming that significant decline begins well only after the time it actually does.
33 Concerns regarding the potential for children born of ARTs to be orphaned are often cited as reasons to limit access for women with potentially terminal diseases. However, some research indicates that stigmatisation leads to a much higher proportion of the population objecting to HIV+ women accessing ARTs, regardless of the real difference in life expectancy compared with women diagnosed with cancer (Evelyn Mok-Lin, et al., 'Public Perceptions of Providing IVF Services to Cancer and HIV Patients,' *Fertility and Sterility* 96, no. 3 [2011]: 724).
34 Daniels would not necessarily object to privately funded ectogenesis for post-menopausal women, on the same grounds that his theory tolerates elective cosmetic surgery for those willing and able to pay for it. A comparable situation would be providing public funding for fertility preservation procedures for cancer patients, but not for those seeking to delay childbearing for social reasons.
35 Smajdor, 'In Defense of Ectogenesis,' 90.
36 Maxwell Mehlman and Karen Visocan, 'Medicare and Medicaid: Are They Just Health Care Systems?' *Houston Law Review* 29, no. 4 (1992): 852.
37 J.C. Patel and M.H. Johnson, 'A Survey of the Effectiveness of the Assessment of the Welfare of the Child in UK In-Vitro Fertilization Units,' *Human Reproduction* 13, no. 3 (1998): 766; Gunsbury, 'Frozen Life's Dominion,' 2205–39; Kathryn Ehrich, et al., 'Social Welfare, Genetic Warfare? Boundary-Work in the IVF/PGD Clinic,' *Social Science & Medicine* 63 (2006): 1213–24.

38 Margaret A. Somerville, 'Birth Technology, Parenting and "Deviance",' *International Journal of Law and Psychiatry* 5 (1982): 123–4.
39 Wayne R. Gillett, et al., 'Development of Clinical Priority Access Criteria for Assisted Reproduction and Its Evaluation on 1386 Infertile Couples in New Zealand,' *Human Reproduction* 27, no. 1 (2011): 131–41; Anita Stuhmucke, 'Lesbian Access to In Vitro Fertilisation,' *Australasian Gay and Lesbian Law Journal* 7 (1997): 15–40; Patel and Johnson, 'A Survey,' 768; Valarie Blake, 'It's an ART Not a Science: State-Mandated Insurance Coverage of Assisted Reproductive Technologies and Legal Implications for Gay and Unmarried Persons,' *Minnesota Journal of Law, Science and Technology* 12, no. 2 (2011): 680.
40 Richard Cookson and Paul Dolan, 'Principles of Justice in Health Care Rationing,' *Journal of Medical Ethics* 26, no. 5 (2000): 326.
41 Jennifer Ruger, *Health and Social Justice* (Oxford: Oxford University Press, 2010), 19; 23.
42 Oommen C. Kurian, 'Rationalising Rationing: The Curious Case of Economic Evaluation in Health,' *Social Scientist* 36, no. 7/8 (2008): 41.
43 Ruger, *Health and Social Justice*, 20.
44 Ibid., 21.
45 John Stuart Mill, *Utilitarianism* (Chicago: Chicago University Press, 1906), 319.
46 Blake, 'It's an ART Not a Science,' 689.
47 Ruger, *Health and Social Justice*, 21.
48 This is currently how access to publicly funded ARTs is managed in New Zealand, leading to the exclusion of a large number of women on grounds such as unhealthy BMI (Gillett, et al. 'Development of Clinical Priority Access Criteria,' 135.)
49 Peter Singer, et al., 'Double Jeopardy and the Use of QALYs in Health Care Allocation,' *Journal of Medical Ethics* 21, no. 3 (1995): 150.
50 Mary Anne Warren, 'The Moral Significance of Birth,' *Hypatia* 4, no. 3 (1989): 50.
51 Lauritzen, *Pursuing Parenthood*, 23.
52 Sen, 'Is Health Care a Luxury?' 151.
53 Saeko Kikuzawa, et al., 'Similar Pressures, Different Contexts: Public Attitudes toward Government Intervention for Health Care in 21 Nations,' *Journal of Health and Social Behavior* 49, no. 4 (2008): 389.
54 Yusuf and Siedlecky, 'Female Sterilizing,' 78.
55 Susan Ettner, et al., 'How Low Birthweight and Gestational Age Contribute to Increased Inpatient Costs for Multiple Births,' *Inquiry* 34, no. 4 (1997): 325.
56 Gillian Harris, et al., '"Seeing the Baby": Pleasures and Dilemmas of Ultrasound Technologies for Primaparous Australian Women,' *Medical Anthropology Quarterly* 18, no. 1 (2004): 24; 27; Folbre, *Who Pays for the Kids?* 104.
57 Pence, 'What's So Good about Natural Motherhood,' 87.

58 L. Hilder, et al. 'Australia's Mothers and Babies 2012,' Perinatal Statistics Series no. 30. Cat. no. PER 69 (Canberra: Australian Institute of Health and Welfare, 2012), 72.
59 Hannah Blencowe, et al., 'Born Too Soon: The Global Epidemiology of 15 Million Preterm Births,' *Reproductive Health* 10, sup. 1 (2013): S2.
60 Sue Dunlevy, 'Growth Spurt in Cost of Birthing,' *Adelaide Advertiser*, 29 January 2015. In this article Medibank Private reveals their highest insurance payout in 2014 was for the treatment of a very premature baby, costing $740,000. They note that of the ten highest payouts for the year, seven were for premature infant care.
61 Loane Skene, 'Recent Developments in Stem Cell Research: Social, Ethical, and Legal Issues for the Future,' *Indiana Journal of Global Legal Studies* 17, no. 2 (2010): 232.
62 Simonstein and Mashiach-Eizenberg, 'The Artificial Womb,' 93.
63 Firestone, *The Dialectic of Sex*, 197.
64 Simonstein and Mashiach-Eizenberg, 'The Artificial Womb,' 93.
65 Sheryl N. Hamilton, 'Traces of the Future: Biotechnology, Science Fiction, and the Media,' *Science Fiction Studies* 30, no. 2 (2003): 267; Sherryl Vint, 'Introduction: Science Fiction and Biopolitics,' *Science Fiction Film and Television* 4, no. 2 (2011): 167.

Conclusion

Abstract: *Kendal discusses areas for future research regarding the ethics of ectogenesis, including the need to consider the potential impact of this technology in the developing world.*

Keywords: developing world bioethics

Kendal, Evie. *Equal Opportunity and the Case for State Sponsored Ectogenesis.* Basingstoke: Palgrave Macmillan, 2015. DOI: 10.1057/9781137549877.0008.

There are various physical, social and economic burdens associated with pregnancy and childbirth that serve to unfairly disadvantage women. As such, there are grounds, both medical and ethical, to fund research into developing an alternative method of reproduction that does not rely on women's personal sacrifices for the good of the species. From both an equal opportunity and feminist perspective it is important to note that in pronatalist societies, including Australia, women's social, educational and occupational development are often de-prioritised as a result of their reproductive capacity. This capacity has historically been used to promote sex-based discrimination in the workplace and still continues to adversely affect women's employment opportunities. The under-representation of women in certain positions of social authority can in part be attributed to the impact of taking time away from paid employment for childbearing purposes, particularly during crucial times in career development. In turn, these negative employment outcomes are generally believed to contribute to reduced fertility rates in many industrialised countries, with professional women often choosing to postpone childbearing or electing to remain childless.

Ectogenesis could provide a means to redress the biological inequality that demands women alone be subjected to the risks associated with pregnancy and childbirth, while society at large reaps the benefits of their reproductive labour. By removing the necessity of physical gestation, greater equality in reproduction could be reached between men and women, between women and other women (including the fertile and infertile) and between heterosexual and homosexual individuals. Ectogenesis would also yield health benefits for the fetus by removing the risk of teratogen exposure in the uncontrolled uterine environment. Although protections would need to be in place to ensure no one is unduly disadvantaged by the advent of this technology, I argue that as long as current pregnancy anti-discrimination policies and abortion laws remain unaffected, and physical pregnancy remains an option for those women wishing to pursue it, the ethical provision of ectogenesis could be achieved. However, given the significant benefits such a technology could yield, I argue it should not be available only to those able to afford it, but rather should be state sponsored through public healthcare systems, such as Australia's Medicare. Furthermore, I argue that there is a responsibility for the state not only to fund the distribution of this technology in the future, but also to actively pursue ectogenesis research in the present, in order to make the potential for equality in reproduction between the sexes a reality.

The major areas for future research on this topic include the need for an in-depth cost–benefit analysis to assess the financial viability of providing public access to this technology and determine the appropriate limitations that should be imposed on its use. Such limitations include whether ectogenesis should be used for the genetic manipulation or enhancement of the fetus, and whether there should be any age limit on who can use this technology to procreate. There are also opportunities to explore cross-cultural attitudes toward ectogenesis between countries, and the impact this technology would have on domestic and international adoption protocols in Australia and overseas. While this book has focused only on industrialised countries, there is also the need for future research to consider how ectogenesis could enhance the lives of women living in the developing world, where the consequences of involuntary childlessness can be life-threatening. Reports coming out of sub-Saharan Africa, for example, indicate that some women experiencing difficulty conceiving are so fearful of being ostracised by their communities or abandoned by their spouses that they are taking their own lives, while others are risking HIV infection to increase their chances of pregnancy through unprotected sex with multiple partners.[1] Despite the fact that the majority of infertile couples in the world reside in developing countries,[2] Giuseppe Benagiano et al. claim that 'Western countries have cultivated the belief that overpopulation, not infertility, is the major problem of developing countries,' thereby ignoring the individual suffering of women and couples within resource-poor countries and their claim to ARTs.[3]

From a feminist perspective, ectogenesis has the potential to greatly change women's position in society and to call for a re-evaluation of cultural beliefs about the nature of family and the traditional gender roles associated with parenthood. As such, I am optimistic that, if handled correctly, the development of this technology could serve to promote greater equality and reproductive liberty for women.

Notes

1 Helen Pilcher, 'Fertility on a Shoestring,' *Nature* 442, no. 31 (2006): 975–7.
2 Yulian Zhao, et al., 'In Vitro Fertilization: Four Decades of Reflection and Promises,' *Biochemica et Biophysica Acta* 1810, no. 9 (2011): 849.
3 Benagiano, et al., 'Robert G Edwards and the Roman Catholic Church,' 671.

Bibliography

Alghrani, Amel and Margaret Brazier. 'What Is It? Whose It? Re-Positioning the Fetus in the Context of Research?' *Cambridge Law Journal* 70, no. 1 (2011): 51–82.

American College of Obstetricians and Gynecologists. 'Pain Relief during Labor.' *Committee Opinion*, no. 295 (2004, reaffirmed 2008): 1.

Anderson, Marilyn. 'Fertility Futures: Implications of National, Pronatalistic Policies for Adolescent Women in Australia,' in *Proceedings of the International Women's Conference, Toowoomba, Queensland*, 26–29 September 2009. Queensland: University of South Queensland, 2007, pp. 40–5.

Andrews, Lori B. 'My Body, My Property.' *The Hastings Center Report* 16, no. 5 (1986): 28–38.

Anolak, Helena. 'Our Bodies, Our Choices: Australian Law on Foetal Personhood.' *Women and Birth* 28, no. 1 (2015): 60–4.

Arditti, Rita. 'Reducing Women to Matter.' *Women's Studies International Forum* 8, no. 6 (1985): 577–82.

Arens, Jutta, Mark Schoberer, Anne Lohr, Thorsten Orlikowsky, Matthias Seehase, Reint K. Jellema, Jennifer J. Collins, Boris W. Kramer, Thomas Schmitz-Rode and Ulrich Steinseifer. 'NeonatOx: A Pumpless Extracorporeal Lung Support for Premature Neonates.' *Artificial Organs* 35, no. 11 (2011): 997–1001.

Åsling-Monemi, Kajsa, Rodolfo Peña, Mary Carroll Ellsberg and Lars Åke Persson. 'Violence Against Women Increases the Risk of Infant and Child

Mortality: A Case-Referent Study in Nicaragua.' *Bulletin of the World Health Organization* 81, no. 1 (2003): 10–18.

Australian Bureau of Statistics. 'Causes of Death, Australia, 2010. Perinatal Deaths: Main Condition in Mother.' Last Modified 14 March 2013. Available at: http://www.abs.gov.au/ausstats/abs@.nsf/Products/C26E274706A50885CA2579C6000F73F1?opendocument.

——. 'The Health and Welfare of Australia's Aboriginal and Torres Strait Islander Peoples, 2008.' Last Modified 22 May 2010. Available at: http://www.aihw.gov.au/WorkArea/DownloadAsset.aspx?id=6442458617.

Australian Government Department of Human Services. 'Paid Parental Leave: Employee Eligibility.' Last Modified 22 January 2015. Available at: http://www.humanservices.gov.au/business/services/centrelink/paid-parental-leave-scheme-for-employers/employee-eligibility.

Australian Government Health Workforce Australia. 'Australia's Health Workforce Series: Doctors in Focus 2012.' Adelaide: Health Workforce Australia, 2012.

Baker, Jeffery P. 'The Incubator and the Medical Discovery of the Premature Infant.' *Journal of Perinatology* 5 (2000): 321–8.

——. 'The Incubator Controversy: Pediatricians and the Origins of Premature Infant Technology in the United States, 1890 to 1910.' *Pediatrics* 87, no. 5 (1991): 654–62.

Bayne, Tim and Avery Kolers. 'Toward a Pluralist Account of Parenthood.' *Bioethics* 17, no. 3 (2003): 221–42.

Benagiano, Giuseppe, Sabina Carrara and Valentina Filippi. 'Robert G Edwards and the Roman Catholic Church.' *Reproductive Biomedicine Online* 22 (2011): 665–72.

Beral, V., P. Doyle, S.L. Tan, B.A. Mason and S. Campbell. 'Outcome of Pregnancies Resulting from Assisted Conception.' *British Medical Bulletin* 46, no. 3 (1990): 753–68.

Biervliet, F.P., S.D. Maguiness, D.M. Hay, S.R. Killick and S.L. Atkin. 'Induction of Lactation in the Intended Mother of a Surrogate Pregnancy: Case Report.' *Human Reproduction* 16, no. 3 (2000): 581–3.

Biggers, John D. 'IVF and Embryo Transfer: Historical Origin and Development.' *Reproductive Biomedicine Online* 25, no. 2 (2012): 118–27.

Binns, Corey. 'The Shark Factory: An Artificial Uterus Gives an Endangered Species a Shot at Survival.' *Popular Science* 275, no. 1 (2009): 34.

Biscoe A. and G. Kidson-Gerber. ' "Avoidable" Death of a Jehovah's Witness with Acute Promyelocytic Leukaemia: Ethical

Considerations and the Internal Conflicts and Challenges Encountered by Practitioners.' *Internal Medicine Journal* 45, no. 4 (2015): 461–2.

Blake, Valarie. 'It's an ART Not a Science: State-Mandated Insurance Coverage of Assisted Reproductive Technologies and Legal Implications for Gay and Unmarried Persons.' *Minnesota Journal of Law, Science and Technology* 12, no. 2 (2011): 651–713.

Blank, Robert. 'Assisted Reproduction and Reproductive Rights: The Case of In Vitro Fertilization.' *Politics and the Life Sciences* 16, no. 2 (1997): 279–88.

Blencowe, Hannah, Simon Cousens, Doris Chou, Mikkel Oestergaard, Lale Say, Ann-Beth Moller, Mary Kinney and Joy Lawn. 'Born Too Soon: The Global Epidemiology of 15 Million Preterm Births.' *Reproductive Health* 10, sup. 1 (2013): S2.

Boivin, Jacky, Janet Takefman and Andrea Braverman. 'The Fertility Quality of Life (FertiQoL) Tool: Development and General Psychometric Properties.' *Human Reproduction* 26, no. 8 (2011): 2084–91.

Boushey, Heather. 'The Role of Government in Work-Family Conflict.' *The Future of Children* 21, no. 2 (2011): 163–90.

Brooks, Michael. 'Faith in Denial.' *New Scientist*, 26 July 2008, 18.

Bruckschwaiger, Otto. 'Four Cases of Advanced and Full-Term Ectopic Pregnancy.' *Canadian Medical Association Journal* 76, no. 9 (1957): 758–61.

Bulletti, Carlo, Antonio Palagiano, Caterina Pace, Angelica Cerni, Andrea Borini and Dominique de Ziegler. 'The Artificial Womb.' *Annals of the New York Academy of Science* 1221 (2011): 124–8.

Callahan, Joan C. and James W. Knight. 'Women, Fetuses, Medicine, and the Law,' in *Feminist Perspectives in Medical Ethics*, eds Helen Bequaert Holmes and Laura M. Purdy, pp. 224–39. Bloomington: Indiana University Press, 1992.

Cameron, C. and R. Williamson, 'Is There an Ethical Difference between Preimplantation Genetic Diagnosis and Abortion?' *Journal of Medical Ethics* 29, no. 2 (2003): 90–2.

Cannold, Leslie. 'Women, Ectogenesis and Ethical Theory.' *Journal of Applied Philosophy* 12, no. 1 (1995): 55–64.

Carter, Lucy. 'Pregnancy Overtakes Disability as Top Source of Workplace Discrimination Complaints.' *ABC News*, 6 November 2013.

Available at: http://www.abc.net.au/news/2013-11-06/pregnancy-overtakes-disability-as-the-top-source-of-discriminati/5072904.
Casas, Nancy. 'Sex Discrimination on the Basis of Pregnancy: Australia's Report on Pregnancy Discrimination Should Make the United States Re-evaluate the Effectiveness of the Pregnancy Discrimination Act in Eliminating Pregnancy Discrimination in the Workplace.' *Transnational Law and Contemporary Problems* 11 (2001): 141–74.
Central Intelligence Agency World Factbook. 'Australia-Oceania.' Last Modified 21 April 2015. Available at: https://www.cia.gov/library/publications/the-world-factbook/geos/as.html.
Chang, M.C. 'My Work on the Transplantation of Mammalian Eggs.' *Theriogenology* 19, no. 3 (1983): 293–303.
Chavatte-Palmer, P., R. Lévy and P. Boileau. 'Une Reproduction Sans Utérus? État des Lieux de L'ectogenèse' [Reproduction without a Uterus? State of the Art of Ectogenesis]. *Gynécologie Obstétrique and Fertilité* 40, no. 11 (2012): 695–7.
Clayton, Jay. 'The Ridicule of Time: Science Fiction, Bioethics, and the Posthuman.' *American Literary History* 25, no. 2 (2013): 317–43.
Cohen, Jonathan. 'The Pregnant Traveller.' *Medicine Today* 9, no. 5 (2008): 65–9.
Colker, Ruth. *Pregnant Men: Practice, Theory, and the Law.* Bloomington: Indiana University Press, 1994.
Colpin, Hilde. 'Parenting and Psychosocial Development of IVF Children: Review of the Research Literature.' *Developmental Review* 22, no. 4 (2002): 644–73.
Conway, David. 'Do Women Benefit from Equal Opportunities Legislation?' in *Equal Opportunities: A Feminist Fallacy*, ed. Caroline Quest, pp. 53–68. London: Goron Pro-Print, 1992.
Cookson, Richard and Paul Dolan. 'Principles of Justice in Health Care Rationing.' *Journal of Medical Ethics* 26, no. 5 (2000): 323–9.
Corderoy, Amy. '"Pregnant Jehovah's Witness" Decision to Refuse Treatment "Harrowing" for Hospital Staff after Mother and Baby Die." *The Sydney Morning Herald*, 6 April 2015. Available at: http://www.smh.com.au/nsw/pregnant-jehovahs-witness-decision-to-refuse-treatment-harrowing-for-hospital-staff-after-mother-and-baby-die-20150406-1mf570.html.
Corea, Genoveffa. 'How the New Reproductive Technologies Could be Used to Apply the Brothel Model of Social Control over Women.' *Women's Studies International Forum* 8, no. 4 (1985): 299–305.

Curtis, Kimberley F. 'Hannah Arendt, Feminist Theorizing, and the Debate over New Reproductive Technologies.' *Polity* 28, no. 2 (1995): 159–87.

Daniels, Norman. *Just Health: Meeting Health Needs Fairly*. Cambridge: Cambridge University Press, 2008.

Dawe, Gavin S., Xiao Wei Tan and Zhi-Cheng Xiao. 'Cell Migration from Baby to Mother.' *Cell Adhesion and Migration* 1, no. 1 (2007): 19–27.

Dixon, Suzanne. 'Conclusion – The Enduring Theme,' in *Stereotypes of Women in Power: Historical Perspective and Revisionist Views*, eds Barbara Garlick, Suzanne Dixon and Pauline Allen, pp. 209–225. New York: Greenwood Press, 1992.

Doherty, S.R. 'Could We Care for Amillia in Rural Australia?' *Rural and Remote Health* 7, no. 4 (2007): 768.

Donchin, Anne. 'The Future of Mothering: Reproductive Technology and Feminist Theory.' *Hypatia* 1, no. 2 (1986): 121–38.

Dubey, Anil K. *Infertility: Diagnosis, Management, & IVF*. New Delhi: Jaypee Brothers Medical Publishers, 2012.

Dunlevy, Sue. 'Growth Spurt in Cost of Birthing.' *Adelaide Advertiser*, 29 January 2015.

Dworkin, Gerald. 'Is More Choice Better than Less?' in *Ethical Principles for Social Policy*, ed. John Howie, pp. 78–96. Carbondale: South Illinois University Press, 1983.

Edwards, John N. 'New Conceptions: Biosocial Innovations and the Family.' *Journal of Marriage and Family* 53, no. 2 (1991): 349–60.

Ehrich, Kathryn, Clare Williams, Rosamund Scott, Jane Sandall and Bobbie Farsides. 'Social Welfare, Genetic Warfare? Boundary-Work in the IVF/PGD Clinic.' *Social Science & Medicine* 63 (2006): 1213–24.

Ellis, Rachel. 'The Botched Epidurals Making Women Terrified of Giving Birth: Is This the Real Reason for Soaring Caesareans?' *The Daily Mail*, 23 November 2010. Available at: http://www.dailymail.co.uk/health/article-1332129/Botched-epidurals-making-women-terrified-giving-birth-Is-real-reason-soaring-Caesareans.html.

Ettner, Susan, Cindy Christiansen, Tamara Callahan and Janet Hall. 'How Low Birthweight and Gestational Age Contribute to Increased Inpatient Costs for Multiple Births.' *Inquiry* 34, no. 4 (1997): 325–39.

Farquhar, C.M., Y.A. Wang and E.A. Sullivan. 'A Comparative Analysis of Assisted Reproductive Technology Cycles in Australia and New Zealand 2004–2007.' *Human Reproduction* 25, no. 9 (2010): 2281–9.

Feldman, Harold. 'A Comparison of Intentional Parents and Intentionally Childless Couples.' *Journal of Marriage and Family* 43, no. 3 (1981): 593–600.

Firestone, Shulamith. *The Dialectic of Sex: The Case for Feminist Revolution*. New York: William Morrow and Company, 1970.

Fischer, Eileen, Cele Otnes and Linda Tuncay. 'Pursuing Parenthood: Integrating Cultural and Cognitive Perspectives on Persistent Goal Striving.' *Journal of Consumer Research* 34, no. 4 (2007): 425–40.

Folbre, Nancy. *Who Pays for the Kids? Gender and the Structures of Constraint*. London: Routledge, 1994.

Ford, Norman M. 'A Catholic Ethical Approach to Human Reproductive Technology.' *Reproductive BioMedicine Online* 17, sup. 3 (2008): 39–48.

Garrido, Nicolás, José Bellver, José Remohí, Carlos Simón and Antonio Pellicer. 'Cumulative Live-Birth Rates Per Total Number of Embryos Needed to Reach Newborn in Consecutive In Vitro Fertilization (IVF) Cycles: A New Approach to Measuring the Likelihood of IVF Success.' *Fertility and Sterility* 96, no. 1 (2011): 40–6.

Gillam, Lynn. 'Prenatal Diagnosis and Discrimination against the Disabled.' *Journal of Medical Ethics* 25, no. 2 (1999): 163–71.

Gillett, Wayne R., John C. Peek and G. Peter Herbison. 'Development of Clinical Priority Access Criteria for Assisted Reproduction and Its Evaluation on 1386 Infertile Couples in New Zealand.' *Human Reproduction* 27, no. 1 (2011): 131–41.

Gimenez, Martha E. 'The Mode of Reproduction in Transition: A Marxist-Feminist Analysis of the Effects of Reproductive Technologies.' *Gender and Society* 5, no. 3 (1991): 334–50.

Golden, Nina G. 'Pregnancy and Maternity Leave: Taking Baby Steps towards Effective Policies.' *Journal of Law and Family Studies* 8 (2008): 1–38.

Grabe, Erin M. 'Gradual Return to Work: Maximizing Benefits to Corporations and Their Caregiver Employees.' *The Journal of Corporation Law* 37, no. 3 (2012): 699–715.

Gray, Gwen. 'Access to Medical Care under Strain: New Pressures in Canada and Australia.' *Journal of Health Politics, Policy and Law* 23, no. 6 (1998): 905–47.

Grobstein, Clifford, Michael Flower and John Mendeloff. 'External Human Fertilization: An Evaluation of Policy.' *Science* 222, no. 4620 (1983): 127–33.

Gunsburg, Samuel A. 'Frozen Life's Dominion: Extending Reproductive Autonomy Rights to In Vitro Fertilization.' *Fordham Law Review* 65 (1996–7): 2205–39.

Hakim, Catherine. *Work-Lifestyle Choices in the 21st Century: Preference Theory*. Oxford: Oxford University Press, 2000.

Halliday, H.L. 'Surfactants: Past, Present and Future.' *Journal of Perinatology* 28, sup. 1 (2008): S47–56.

Hamilton, Sheryl N. 'Traces of the Future: Biotechnology, Science Fiction, and the Media.' *Science Fiction Studies* 30, no. 2 (2003): 267–82.

Harris, John. 'One Principle and Three Fallacies of Disability Studies.' *Journal of Medical Ethics* 27, no. 6 (2001): 383–7.

Harris, Gillian, Linda Connor, Andrew Bisits and Nick Higginbotham. '"Seeing the Baby": Pleasures and Dilemmas of Ultrasound Technologies for Primaparous Australian Women.' *Medical Anthropology Quarterly* 18, no. 1 (2004): 23–47.

Hayes, Jeanne. 'Female Infertility in the Workplace: Understanding the Scope of the Pregnancy Discrimination Act.' *Connecticut Law Review* 42, no. 4 (2010): 1299–1335.

Heitlinger, Alena. 'Pronatalism and Women's Equality Policies.' *European Journal of Population* 7, no. 4 (1991): 343–75.

Heitman, Elizabeth. 'Social and Ethical Implications of *In Vitro* Fertilization.' *International Journal of Technology Assessment in Health Care* 15, no. 1 (1999): 22–35.

Henry, Barbara, Shane Houston and Gavin Mooney. 'Institutional Racism in Australian Healthcare: A Plea for Decency.' *Medical Journal of Australia* 180, no. 10 (2004): 517–20.

Herbert, Danielle, Jayne Lucke and Annette Dobson. 'Early Users of Fertility Treatment with Hormones and IVF: Women Who Live in Major Cities and Have Private Health Insurance.' *Australian and New Zealand Journal of Public Health* 34, no. 6 (2010): 629–34.

Heriot, M. Jean. 'Fetal Rights versus the Female Body: Contested Domains.' *Medical Anthropology Quarterly* 10, no. 2 (1996): 176–94.

Hermanides, J., M.W. Hollmann, M.F. Stevens and P. Lirk. 'Failed Epidural: Causes and Management.' *British Journal of Anaesthesia* 109, no. 2 (2012): 144–54.

Hewson, Barbara. 'Reproductive Autonomy and the Ethics of Abortion.' *Journal of Medical Ethics* 27, sup. 2 (2001): ii10–ii14.

Hilder, L., Z. Zhichao, M. Parker, S. Jahan and G.M. Chambers. 'Australia's Mothers and Babies 2012.' Perinatal Statistics Series no. 30. Cat. no. PER 69. Canberra: Australian Institute of Health and Welfare, 2012.

Hill, Alexandria J., Kristi R. Van Winden and Curtis R. Cook. 'A True Cornual (Interstitial) Pregnancy Resulting in a Viable Fetus.' *Obstetrics and Gynecology* 121, no. 2 (2013): 427–30.

Hirshman, Linda R. *Get to Work: A Manifesto for Women of the World*. New York: Viking, 2006.

Horowitz, Eran R., Yariv Yogev, Avi Ben-Haroush and Boris Kaplan. 'Women's Attitudes toward Analgesia during Labor – A Comparison between 1995 and 2001.' *European Journal of Obstetrics & Gynecology and Reproductive Biology* 117, no. 1 (2004): 30–2.

Human Rights and Equal Opportunity Commission. 'Pregnant and Productive: It's a Right Not a Privilege to Work While Pregnant. Report of the National Pregnancy and Work Inquiry.' (1999): i–xxi.

Hunt, Sheila C. and Ann M. Martin. *Pregnant Women Violent Men: What Midwives Need to Know*. Oxford: Butterworth-Heinemann, 2001.

Inhorn, Marcia C. 'Making Muslim Babies: Ivf and Gamete Donation in Sunni versus Shi'a Islam.' *Culture, Medicine and Psychiatry* 30, no. 4 (2006): 427–50.

Isaacs, D. 'Moral Status of the Fetus: Fetal Rights If Maternal Autonomy?' *Journal of Paediatrics and Child Health* 39, no. 1 (2003): 58–9.

Ishii, Nozomi, Yumi Kono, Naohiro Yonemoto, Satoshi Kusuda and Masanori Fujimura. 'Outcomes of Infants Born at 22 and 23 Weeks' Gestation.' *Pediatrics* 132, no, 1 (2013): 62–71.

Johanson, Richard, Mary Newburn and Alison MacFarlane. 'Has the Medicalization of Childbirth Gone Too Far?' *British Medical Journal* 324, no. 7342 (2002): 892–5.

Johnson, Harriet McBryde. 'Unspeakable Conversations or How I Spent One Day as a Token Cripple at Princeton University.' *The New York Times Magazine*, 16 February 2003, 50.

Johnstone, Megan-Jane. 'Ethics and Ectogenesis.' *Australian Nursing Journal* 17, no. 11 (2010): 33.

Jones, D. Gareth. *Brave New People: Ethical Issues at the Commencement of Life*. Leicester: Inter-Varsity Press, 1984.

Josefson, Deborah. 'Scientists Fertilise Mouse Eggs without Sperm.' *British Medical Journal* 323, no. 7305 (2001): 127.

Kanazawa, Satoshi. 'Intelligence and Childlessness.' *Social Science Research* 48 (2014): 157–70.

Kendall, Patricia, Lydia C. Medeiros, Virginia Hillers, Gang Chen and Steve DiMascola. 'Food Handling Behaviors of Special Importance for Pregnant Women, Infants and Young Children, the Elderly, and Immune-Compromised People.' *Journal of the American Dietetics Association* 103, no. 12 (2003): 1646–9.

Kidd, Michael, Ian Watts and Deborah Saltman. 'Primary Health Care Reform: Equity Is the Key.' *Medical Journal of Australia* 189, no. 4 (2008): 221–2.

Kikuzawa, Saeko, Sigrun Olafsdottir and Bernice Pescosolido. 'Similar Pressures, Different Contexts: Public Attitudes toward Government Intervention for Health Care in 21 Nations.' *Journal of Health and Social Behavior* 49, no. 4 (2008): 385–99.

Klock, Susan Caruso and Dorothy A. Greenfield. 'Psychological Status of In Vitro Fertilization Patients during Pregnancy: A Longitudinal Study.' *Fertility and Sterility* 73, no. 6 (2000): 1159–64.

Kovacs, Gabor T., Sue A. Breheny and Melinda J. Dear. 'Embryo Donation at an Australian University In-Vitro Fertilisation Clinic: Issues and Outcomes.' *Medical Journal of Australia* 178, no. 3 (2003): 127–9.

Krauss, Lawrence. 'Science the Catholic Church Can't Ignore.' *New Scientist*, 7 February 2009, 25.

Kurian, Oommen C. 'Rationalising Rationing: The Curious Case of Economic Evaluation in Health.' *Social Scientist* 36, no. 7/8 (2008): 37–63.

Ladd, Rosalind Ekman. 'Women in Labor: Some Issues about Informed Consent,' in *Feminist Perspectives in Medical Ethics*, eds Helen Bequaert Holmes and Laura M. Purdy, pp. 216–23. Bloomington: Indiana University Press, 1992.

Lairson, David, Paul Hindson and Alan Hauquitz. 'Equity of Heath Care in Australia.' *Social Science Medicine* 41, no. 4 (1995): 475–82.

Lambert, Helen H. 'Biology and Equality: A Perspective on Sex Differences.' *Signs* 4, no. 1 (1978): 97–117.

Lasker, Judith. *In Search of Parenthood: Coping with Infertility and High-Tech Conception*. Philadelphia: Temple University Press, 1994.

Lauritzen, Paul. 'Catholics & IVF: The Next Big Battleground?' *Commonweal* 12 August 2005, 10.

——. *Pursuing Parenthood*. Indiana: Indiana University Press, 1993.

Lavender. 'Lesbian Feminist Activism in Australia.' *Off Our Backs* 36, no. 3 (2006): 71–4.

Leeder, Stephen R. 'Achieving Equity in the Australian Healthcare System.' *Medical Journal of Australia* 179, no. 9 (2003): 475–8.

Leeners, Brigitte, Kirsten Geraedts, Bruno Imthurn and Ruth Stiller. 'The Relevance of Age in Female Human Reproduction – Current Situation in Switzerland and Pathophysiological Background from a Comparative Perspective.' *General and Comparative Endocrinology* 188 (2013): 166–74.

Lockwood, Gillian. 'Pregnancy, Autonomy and Paternalism.' *Journal of Medical Ethics* 25, no. 6 (1999): 537–40.

Lorber, Judith. 'Choice, Gift, or Patriarchal Bargain? Women's Consent to *In Vitro* Fertilization in Male Infertility,' in *Feminist Perspectives in Medical Ethics*, eds Helen Bequaert Holmes and Laura M. Purdy, pp. 169–80. Bloomington: Indiana University Press, 1992.

Losco, Joseph and Mark Shublak. 'Paternal-Fetal Conflict: An Examination of Paternal Responsibilities to the Fetus.' *Politics and the Life Sciences* 13, no. 1 (1994): 63–75.

Lynch, Michael. 'Forgotten Fathers.' *The Body Politic: Gay Liberation Journal*, April 1978.

Mac Dougall, K., Y. Beyene and R.D. Nachtigall. 'Age Shock: Misperceptions of the Impact of Age on Fertility before and after IVF in Women Who Conceived after Age 40.' *Human Reproduction* 28, no. 2 (2013): 350–6.

Majumdar, Debarun. 'Choosing Childlessness: Intentions of Voluntary Childlessness in the United States.' *Michigan Sociological Review* 18 (2004): 108–35.

Makdisi, June Mary Zekan. 'Genetically Correct: The Political Use of Reproductive Technology.' *Pepperdine Law Review* 32, no. 1 (2004–2005): 1–38.

Mallia, Pierre. 'Problems Faced with Legislating for IVF Technology in a Roman Catholic Country.' *Medicine, Health Care and Philosophy* 13 (2010): 77–87.

Martin, Richard J., Avroy A. Fanaroff and Michelle C. Walsh. *Fanaroff and Martin's Neonatal-Perinatal Medicine: Diseases of the Fetus and Infant*, 10th ed. St Louis: Elsevier, 2015.

Mason, Chris. 'Making People: Today's Wariness of Reproductive Technologies Stems from Myths, Legends and Hollywood.' *Nature* 471, no. 7338 (2011): 297–9.

McDonald, Paula, Kerriann Dear and Sandra Backstrom. 'Expecting the Worst: Circumstances Surrounding Pregnancy Discrimination at Work and Progress to Formal Redress.' *Industrial Relations Journal* 39, no. 3 (2008): 229–47.

McLeod, Carolyn and Julie Ponesse. 'Infertility and Moral Luck: The Politics of Women Blaming Themselves for Infertility.' *International Journal of Feminist Approaches to Bioethics* 1, no. 1 (2008): 126–44.

McMahan, Jeff. 'Causing Disabled People to Exist and Causing People to be Disabled.' *Ethics* 116, no. 1 (2005): 77–99.

McMahon, Catherine A., Judy A. Ungerer, Christopher Tennant and Douglas Saunders. 'Psychosocial Adjustment and the Quality of the Mother-child Relationship at Four Months Postpartum after Conception by In Vitro Fertilization.' *Fertility and Sterility* 68, no. 3 (1997): 492–500.

Medew, Julia. 'Surrogacy for Gay Couples in Victoria, Australia.' *The Sydney Morning Herald*, 7 March 2015. Available at: http://www.smh.com.au/national/health/surrogacy-for-gay-couples-in-victoria-australia-20150306-13xcd5.html.

Mehlman, Maxwell and Karen Visocan. 'Medicare and Medicaid: Are They Just Health Care Systems?' *Houston Law Review* 29, no. 4 (1992): 835–65.

Meyers, Diana. 'The Rush to Motherhood: Pronatalist Discourse and Women's Autonomy.' *Signs* 26, no. 3 (2001): 735–73.

Mill, John Stuart. *Utilitarianism*. Chicago: Chicago University Press, 1906.

Mok-Lin, Evelyn, Stacey Missmer, Katharine Berry, Lisa Soleymani Lehmann and Elizabeth S. Ginsburg. 'Public Perceptions of Providing IVF Services to Cancer and HIV Patients.' *Fertility and Sterility* 96, no. 3 (2011): 722–7.

Murphy, Julien S. 'Is Pregnancy Necessary? Feminist Concerns about Ectogenesis.' *Hypatia* 4, no. 3 (1989): 66–84.

Murphy, Timothy F. 'The Ethics of Impossible and Possible Changes to Human Nature.' *Bioethics* 26, no. 4 (2012): 191–7.

Nozick, Robert. *Anarchy, State, and Utopia*. New York: Basic Books, 1974.

Oakley, Ann. 'Gender and Generation: The Life and Times of Adam and Eve,' in *Women and the Life Cycle: Transitions and Turning-Points*, eds Patricia Allatt, Teresa Keil, Alan Bryman and Bill Bytheway, pp. 13–32. Essex: Macmillan Press, 1987.

O'Boyle, Amy L., Gary D. Davis and Byron C. Calhoun. 'Informed Consent and Birth: Protecting the Pelvic Floor and Ourselves.' *American Journal of Obstetrics and Gynecology* 187, no. 4 (2002): 981–3.

Oliver, S., L. Rajan, H. Turner, A. Oakley, V. Entwhistle, I. Watt, T. A. Sheldon, J. Rosser, M. Newburn, M. Gready, Paul Chamberlain and P.A. Boyd. 'Informed Choice for Users of Health Services: Views on Ultrasonography Leaflets of Women in Early Pregnancy, Midwives, and Ultrasonographers.' *British Medical Journal* 313, no. 7067 (1996): 1251–5.

Otway, Nick and Megan Ellis. 'Construction and Test of an Artificial Uterus for Ex Situ Development of Shark Embryos.' *Zoo Biology* 31, no. 2 (2012): 197–205.

Pagidas, K., C. Nezhat, S. Carson, A. Nezhat, M. Cesario, S. Woodard, A. Nadal and J.E. Buster. 'Previvo Uterine Lavage Catheter: A Novel Device for the Recovery of In Vivo Derived Human Embryos by Non-Surgical Uterine Lavage.' *Fertility and Sterility* 102, no. 3, sup. (2014): e33.

Parliament of Australia. *Prohibition of Human Cloning for Reproduction and the Regulation of Human Embryo Research Amendment Act 2006*, no. 172 (2006): 1–33.

Patel J.C. and M.H. Johnson. 'A Survey of the Effectiveness of the Assessment of the Welfare of the Child in UK In-Vitro Fertilization Units.' *Human Reproduction* 13, no. 3 (1998): 766–70.

Patterson, Lou Ann and John Defrain. 'Pronatalism in High School Family Studies Texts.' *Family Relations* 30, no. 2 (1981): 211–17.

Payne, Catherine D. 'Stem Cell Research and Cloning for Human Reproduction: An Analysis of the Laws, the Direction in Which They May Be Heading in Light of Recent Developments, and Potential Constitutional Issues.' *Mercer Law Review* 61 (2010): 943–76.

Pence, Gregory. 'What's So Good about Natural Motherhood? (In Praise of Unnatural Gestation),' in *Ectogenesis: Artificial Womb Technology and the Future of Human* Reproduction, eds Scott Gelfand and John Shook, pp. 77–88. New York: Rodopi, 2006.

Pilcher, Helen. 'Fertility on a Shoestring.' *Nature* 442, no. 31 (2006): 975–7.

Pollitt, Katha. *Pro: Reclaiming Abortion Rights*. New York: Picador, 2014.

Powers, Madison. 'Privacy and Genetics,' in *A Companion to Genethics*, eds Justine Burley and John Harris, pp. 364–77. Oxford: Blackwell, 2002.

Ragu, Tonse N.K. 'An Extant Hess Incubator on Display,' [letter to the editor]. *Pediatrics* 107, no. 4 (2001): 805.

Raymond, Janice G. 'Reproductive Gifts and Gift Giving: The Altruistic Woman,' in *Life Choices: A Hastings Center Introduction to Bioethics,*

2nd ed., eds Joseph H. Howell and William Frederick Sale, pp. 395–406. Washington D.C.: Georgetown University Press, 2000.

Reinhardt, Uwe. 'Resource Allocation in Health Care: The Allocation of Lifestyles to Providers.' *The Milbank Quarterly* 65, no. 2 (1987): 153–76.

Reuben, David. 'Organizational Interventions to Improve Health Outcomes of Older Persons.' *Medical Care* 40, no. 5 (2002): 416–28.

Rice, Nigel and Peter C. Smith. 'Ethics and Geographical Equity in Health Care.' *Journal of Medical Ethics* 27, no. 4 (2001): 256–61.

Roberts, Dorothy E. 'The Genetic Tie.' *The University of Chicago Law Review* 62, no. 1 (1995): 209–73.

Roberts, Elizabeth F.S. 'God's Laboratory: Religious Rationalities and Modernity in Ecuadorian In Vitro Fertilization.' *Culture, Medicine and Psychiatry* 30, no. 4 (2006): 507–36.

Roberts, Celia and Karen Throsby. 'Paid to Share: IVF Patients, Eggs and Stem Cell Research.' *Social Science & Medicine* 66 (2008): 159–69.

Ross, Steven L. 'Abortion and the Death of the Fetus.' *Philosophy and Public Affairs* 11, no. 3 (1982): 232–45.

Rothman, Barbara Katz. *The Tentative Pregnancy: How Amniocentesis Changes the Experience of Motherhood.* New York: W. W. Norton & Company, 1993.

Rowland, Robyn. 'A Child at Any Price? An Overview of Issues in the Use of the New Reproductive Technologies, and the Threat to Women.' *Women's Studies International Forum* 8, no. 6 (1985): 539–46.

——. 'Technology and Motherhood: Reproductive Choice Reconsidered.' *Signs* 12, no. 3 (1987): 512–28.

Ruger, Jennifer. *Health and Social Justice.* Oxford: Oxford University Press, 2010.

Sander-Staudt, Maureen. 'Of Machine Born? A Feminist Assessment of Ectogenesis and Artificial Wombs,' in *Ectogenesis: Artificial Womb Technology and the Future of Human Reproduction*, eds Scott Gelfand and John Shook, pp. 109–28. New York: Rodopi, 2006.

Savulescu, Julian. 'Procreative Beneficence: Why We Should Select the Best Children.' *Bioethics* 15, nos. 5–6 (2001): 413–26.

Schaeffer, Pamela. 'In Vitro Fertilization Widely Used.' *National Catholic Reporter*, 15 October 1999, 13–16.

Schenker, J.G. '*In Vitro* Fertilization and Embryo Transfer: Jewish Ethical and Legal Aspects,' in *In Vitro Fertilization, Embryo Transfer and Early Pregnancy*, eds R.F. Harrison, J. Bonnar and W. Thompson, pp. 181–90. Lancaster: MTP Press Limited, 1984.

Schotsmans, Paul T. 'In Vitro Fertilisation: The Ethics of Illicitness? A Personalist Catholic Approach.' *European Journal of Obstetrics & Gynecology and Reproductive Biology* 81 (1998): 235–41.

Schultz, Jessica, H. 'Development of Ectogenesis: How Will Artificial Wombs Affect the Legal Status of a Fetus or Embryo?' *Chicago-Kent Law Review* 84 (2010): 877–906.

Sen, Anindya. 'Is Health Care a Luxury? New Evidence from OECD Data.' *International Journal of Health Care Finance and Economics* 5, no. 2 (2005): 147–64.

Shabalala, Nombuso. 'A New World Court to Judge Gender-Based War Crimes.' *Agenda: Empowering Women for Gender Equity* 52 (2002): 89–90.

Sharma, R.K., T. Said and A. Agarwal. 'Sperm DNA and Its Clinical Relevance in Assessing Reproductive Outcomes.' *Asian Journal of Andrology* 6, no. 2 (2004): 139–48.

Silverio, Maili Malin and Elina Hemminki. 'Practice of *In-vitro* Fertilization: A Case Study from Finland.' *Social Science and Medicine* 42, no. 7 (1996): 975–83.

Silverman, William A. 'Incubator-Baby Side Shows.' *Pediatrics* 64, no. 2 (1979): 127–41.

Simonstein, Frida and Michal Mashiach-Eizenberg. 'The Artificial Womb: A Pilot Study Considering People's Views on the Artificial Womb and Ectogenesis in Israel.' *Cambridge Quarterly of Healthcare Ethics* 18 (2009): 87–94.

Singer, Peter. 'Technology and Procreation: How Far Should We Go?' *Technology Review* 88 (1985): 22–30.

Singer, Peter and Deane Wells. 'Ectogenesis,' in *Ectogenesis: Artificial Womb Technology and the Future of Human Reproduction*, eds Scott Gelfand and John Shook, pp. 9–26. New York: Rodopi, 2006.

———. *Making Babies: The New Science and Ethics of Conception.* New York: Charles Scribner's Sons, 1985.

Singer, Peter, John McKie, Helga Kuhse and Jeff Richardson. 'Double Jeopardy and the Use of QALYs in Health Care Allocation.' *Journal of Medical Ethics* 21, no. 3 (1995): 144–50.

Skene, Loane. 'Recent Developments in Stem Cell Research: Social, Ethical, and Legal Issues for the Future.' *Indiana Journal of Global Legal Studies* 17, no. 2 (2010): 211–44.

Smajdor, Anna. 'In Defense of Ectogenesis.' *Cambridge Quarterly of Healthcare Ethics* 21, no. 1 (2012): 90–103.

——. 'The Moral Imperative for Ectogenesis.' *Cambridge Quarterly of Healthcare Ethics* 16, no. 3 (2007): 336–45.
Somerville Margaret A. 'Birth Technology, Parenting and "Deviance".' *International Journal of Law and Psychiatry* 5 (1982): 123–53.
Souza, João, Antonio Oliveria-Neto, Fernanda Garanhani Surita, José Cecatti, Eliana Amaral and João L. Pinto e Silva. 'The Prolongation of Somatic Support in a Pregnant Woman with Brain-death: A Case Report.' *Reproductive Health* 3, no. 1 (2006): 3.
Sparrow, Robert. 'Is it "Every Man's Right to Have Babies If He Wants Them"? Male Pregnancy and the Limits of Reproductive Liberty.' *Kennedy Institute of Ethics Journal* 18, no. 3 (2008): 275–99.
Spinelli, Margaret G. 'Postpartum Psychosis: Detection of Risk and Management.' *American Journal of Psychiatry* 166, no. 4 (2009): 405–8.
Squier, Susan M. *Babies in Bottles: Twentieth-Century Visions of Reproductive Technology*. New Brunswick: Rutgers University Press, 1994.
Steiger, Eric. 'Not of Woman Born: How Ectogenesis Will Change the Way We View Viability, Birth, and the Status of the Unborn.' *Journal of Law and Health* 23, no. 2 (2010): 143–71.
Stellman, Jeanne and Mary Sue Henifin. 'No Fertile Women Need Apply: Employment Discrimination and Reproductive Hazards in the Workplace,' in *Biological Woman – A Convenient Myth: A Collection of Feminist Essays and a Comprehensive Bibliography*, eds Ruth Hubbard, Mary Sue Henifin and Barbara Fried, pp. 117–45. Cambridge: Schenkman Publishing, 1982.
Stuhmucke, Anita. 'Lesbian Access to In Vitro Fertilisation.' *Australasian Gay and Lesbian Law Journal* 7 (1997): 15–40.
Sutton, Leslie. 'Fetal Surgery for Neural Tube Defects.' *Best Practice and Research Clinical Obstetrics and Gynaecology* 22, no. 1 (2008): 175–88.
Taylor, Lenore. 'Human Rights Commission President Gillian Triggs Hits Back at Critics.' *The Guardian*, 1 April 2015. Available at: http://www.theguardian.com/australia-news/2015/apr/01/human-rights-commission-president-gillian-triggs-hits-back-at-the-critics.
Tong, Rosemary. *Feminist Thought: A Comprehensive Introduction*. Boulder, Colorado, USA: Westview Press, 1989.
Tonts, Matthew and Ann-Claire Larsen. 'Rural Disadvantage in Australia: A Human Rights Perspective.' *Geography* 87, no. 2 (2002): 132–41.
Tranter, Kieran. 'The Speculative Jurisdiction: The Science Fictionality of Law and Technology.' *Griffith Law Review* 20, no. 4 (2011): 817–50.

Tsoumpou, Ioanna, Ahmed M. Mohamed, Clare Tower, Stephen A. Roberts and Luciano G. Nardo. 'Failed IVF Cycles and the Risk of Subsequent Preeclampsia or Fetal Growth Restriction: A Case-Control Exploratory Study.' *Fertility and Sterility* 95, no. 3 (2011): 973–8.

Unno, Nobuya, Yoshinori Kuwabara, Takashi Okai, Koichiro Kido, Hirotoshi Nakayama, Akihiko Kikuchi, Yumiko Narumiya, Shiro Kozuma, Yuji Taketani and Masanori Tamura. 'Development of an Artificial Placenta: Survival of Isolated Goat Fetuses for Three Weeks with Umbilical Arteriovenous Extracorporeal Membrane Oxygenation.' *Artificial Organs* 17, no. 12 (1993): 996–1003.

Van Doorslaer, Eddy, Philip Clarke, Elizabeth Savage and Jane Hall. 'Horizontal Inequities in Australia's Mixed Public/Private Health Care System.' *Health Policy* 86, no. 1 (2008): 97–108.

Van Gelder, Marleen M.H.J., Iris A.L.M. van Rooij, Richard K. Miller, Gerhard A. Zielhuis, Lolkje T.W. de Jong-van den Berg and Nel Roeleveld. 'Teratogenic Mechanisms of Medical Drugs.' *Human Reproduction Update* 16, no. 4 (2010): 378–94.

Vint, Sherryl. 'Introduction: Science Fiction and Biopolitics.' *Science Fiction Film and Television* 4, no. 2 (2011): 161–72.

Warren, Mary Anne. 'Making Babies: The New Science and Ethics of Conception by Peter Singer; Deane Wells.' *Ethics* 97, no. 1 (1986): 288–9.

——. 'The Moral Significance of Birth.' *Hypatia* 4, no. 3 (1989): 46–65.

Williams, Patrick. 'Fiancé of Kymberlie Shepherd, Who Died during Childbirth, Calls for More Research into Rare Amniotic Fluid Embolism.' *ABC News*, 31 October 2014. Available at: http://www.abc.net.au/news/2014-10-31/queensland-mother-dies-from-rare-amniotic-fluid-embolism/5844512.

Wilson, J. 'Not So Special after All? Daniels and the Social Determinants of Health.' *Journal of Medical Ethics* 35, no. 1 (2009): 3–6.

Wolfe, Alexander D. 'Wrongful Selection: Assisted Reproductive Technologies, Intentional Diminishment, and the Procreative Right.' *Thomas M. Cooley Law Review* 25, no. 3 (2008): 475–502.

Woolf, Nicky. 'Purvi Patel Found Guilty of Feticide and Child Neglect over Unborn Baby's Death.' *The Guardian*, 5 February 2015. Available at: http://www.theguardian.com/us-news/2015/feb/04/purvi-patel-found-guilty-feticide-unborn-childs-death.

World Health Organisation Department of Reproductive Health and Research. *Managing Complications in Pregnancy and Childbirth: A Guide for Midwives and Doctors.* Geneva: World Health Organisation, 2007.

Yusuf, Farhat and Stefania Siedlecky. 'Female Sterilizing Operations in New South Wales: A Demographic Perspective.' *Journal of the Australian Population Association* 15, no. 1 (1998): 69–79.

Zhao, Yulian, Paul Brezina, Chao-Chin Hsu, Jairo Garcia, Peter R., Brinsden and Edward Wallach. 'In Vitro Fertilization: Four Decades of Reflection and Promises.' *Biochemica et Biophysica Acta* 1810, no. 9 (2011): 843–52.

Zimmerman, Jan. 'Technology and the Future of Women: Haven't We Met Somewhere Before?' *Women's Studies International Forum* 4, no. 3 (1981): 355–67.

Index

artificial placenta, 34, 35
autonomy, 10, 11, 12, 13, 69, 70, 71, 72, 73, 74

bodily integrity, 6, 47, 48, 49, 70, 73, 74, 75, 76

ectogenesis
   definition of, 2
   impact on abortion, 36, 52, 73, 74, 75, 76
   in animals, 33, 34

fetal harm
   abuse, 48, 49
   experimentation, 79
   teratogens, 6, 18, 57, 105, 113
Firestone, Shulamith, 8, 19, 23, 25, 31, 45, 46, 50, 58, 59, 107, 111, 120

humidicribs, 27

IVF, 28, 51, 69
   history of, 20, 28

Lauritzen, Paul, 39, 60, 69, 70, 89, 104, 110, 123

medicare, 21, 94, 95, 96, 97, 98, 104, 106, 109, 113, 125

mortality
   infant, 27, 97
   maternal, 3, 4, 7

Pareto optimality, 63, 67, 72, 75, 77, 87
pregnancy
   discrimination, 16, 17, 55, 57, 64, 65, 66, 113
   illnesses of, 3, 4, 6
privacy, 9, 12, 13, 47, 49, 57, 73
pronatalism, 9, 10, 52, 70
   definition of, 9

Rowland, Robyn, 25, 30, 40, 51, 60, 91, 127

Singer, Peter, 20, 22, 25, 37, 42, 47, 49, 59, 70, 79, 89, 91, 110, 128, 130, 131
Smajdor, Anna, 3, 22, 31, 44, 45, 50, 53, 58, 59, 60, 81, 87, 91, 92, 100, 109, 128, 131
social infertility, 53, 54, 101
surrogacy, 2, 48, 53, 55, 96

utilitarianism, 101, 102, 103

The manufacturer's authorised representative in the EU is Springer Nature Customer Service Centre GmbH, Europaplatz 3, 69115 Heidelberg, Germany. If you have any concerns regarding our products, please contact ProductSafety@springernature.com

Printed and bound by CPI Group (UK) Ltd, Croydon, CR0 4YY

24/03/2026

02077827-0002